SpringerBriefs in Computer Science

For further volumes:
http://www.springer.com/series/10028

Dongmei Zhao

Power Distribution and Performance Analysis for Wireless Communication Networks

 Springer

Dongmei Zhao
McMaster University
Hamilton, Ontario, Canada
dzhao@mcmaster.ca

ISSN 2191-5768   e-ISSN 2191-5776
ISBN 978-1-4614-3283-8   e-ISBN 978-1-4614-3284-5
DOI 10.1007/978-1-4614-3284-5
Springer New York Dordrecht Heidelberg London

Library of Congress Control Number: 2012932310

Printed on acid-free paper

Springer is part of Springer Science+Business Media (www.springer.com)

To my parents

Contents

Acronyms

AF	Amplify-and-forward
ARQ	Automatic Repeat-reQuest
AWGM	Additive white Gaussian noise
BER	Bit error rate
BS	Base station
CDMA	Code division multiple access
CR	Cognitive radio
CRN	Cognitive radio network
DF	Decode-and-forward
ERA	Equal rate allocation
ELG	Effective link gain
GP	Geometric programming
GSM	Global system for mobile communications
HHO	Hard handoff
KKT	Karush-Kuhn-Tucker
MRC	Maximum ratio combining
MS	Mobile station
OFDMA	Orthogonal frequency division multiple access
p2s	Primary-to-secondary
PRA	Proportional fair rate allocation
QoS	Quality of service
RRM	Radio resource management
RS	Relay station
s2p	Secondary-to-primary
S-AF	Selection amplify-and-forward
S-D	Source-to-destination
SHO	Soft handoff
SINR	Signal-to-interference-plus-noise ratio
TDMA	Time division multiple access
WiMAX	Worldwide Interoperability for Microwave Access
WLAN	Wireless local area network

Chapter 1
Introduction

Abstract This chapter includes two sections. The first section briefly introduces radio resource management (RRM), including RRM problems in the first and second generation wireless cellular communication networks, unique features in wireless communication networks that bring both challenges and opportunities for RRM, and basic functions of RRM in current and future wireless communication networks. The second section focuses on power control and interference management in wireless networks, where the importance of power control and its relationship to other network functions are first introduced, then power allocation problems are formulated mathematically, and finally existence of feasible solutions and other properties related to power allocations are analyzed.

Keywords: Radio resource management, power control, signal-to-interference-plus-noise ratio.

1.1 Radio Resource Management

1.1.1 Unique Features in Wireless Communication Networks

Radio resource management (RRM) generally refers to how different types of radio resources in wireless networks are shared among mobile users or transmission links. The basic objectives for RRM are two folds: i) to satisfy the quality-of-service (QoS) requirements of the users, and ii) to efficiently utilize the radio resources. These objectives are basically the same as that for resource management in wired networks. However, there are some unique features in wireless networks that make the resource management problem fundamentally different from that in wired networks. Although most of these features were considered to affect the wireless network performance in negative ways, they have been exploited in different ways in modern wireless communication networks for improving the network performance, depending on specific applications and network scenarios. Some of these features are listed below:

- First, the radio propagation channel can experience random fading, which makes the data transmissions prone to transmission errors. As a result, the amount of resources allocated to each user can be significantly different depending on the user's channel conditions. Users in deeper fading require more radio resources, such as higher transmission power or wider bandwidth, in order to combat the poor channel conditions so that they can receive the same signal quality as other users with better channel conditions. When a large number of users are in deep fading, the system capacity can be reduced significantly. On the other hand, the random fading may result in channel conditions much better than average at times and locations, and this provides opportunities for improving QoS to the users and efficiency of resource utilization. If users can delay their transmissions and wait until their channel conditions become good, the network resources can be more efficiently utilized, and the system capacity can be improved. This type of ideas have been widely used, especially for serving elastic traffic, which does not have strict latency requirements.
- Second, due to the broadcast nature of the radio channel, one user's transmission can reach all the other users nearby and interfere their communications, if these users happen to receive at the same frequency channel at the same time. Such co-channel interference reduces the QoS of the users. In order to prevent strong co-channel interference, transmission time and power of the users should be carefully scheduled. Users within each other's interference range should avoid to transmit at the same time. As a result, the per-user throughput can be low in areas of high user density. On the other hand, since one transmission can simultaneously reach multiple destinations, the broadcast feature has been used in various ways to save network resources and improve the network performance. For example, broadcasting facilitates peer mobile stations (MSs) to relay traffic for each other, which helps extend communication range, improve transmission quality, and increase system capacity. Since the relay request of one MS may be received by multiple peer MSs, the best relay can be selected from the peers to optimize the system performance.
- Furthermore, MSs can change their locations from time to time. Such location changes can affect the connectivity between the communicating parties, change the amount of required resources for the communications, and further affect the mutual interference conditions among the users. This effect complicates routing, QoS provisioning, and performance analysis. In order to make accurate resource allocations and guarantee the QoS of the users, the network may need to track the users' current locations and predict their future locations, which can be difficult and sometimes is impossible. On the other hand, the location changes provide spatial diversity, which can be exploited in RRM to save the network resources or provide the users with better QoS [1, 2, 3]. For example, when all users are fixed and have different channel conditions, the users will receive different QoS, if each is assigned an equal amount of resources; alternatively, if they are served with the same QoS, a different amount of resources is consumed by each user.

In this example, the first service mode results in fair QoS but unfair resource allocations, while the second results in fair resource allocations but unfair QoS. Achieving both fair QoS and fair resource allocations is unlikely in such a scenario. If the users are moving, the mobility may automatically bring some levels of fairness, in terms of both QoS and resource usage, over a long term as they each go through different locations and experience different channel and interference conditions.

1.1.2 RRM in Different Wireless Communication Networks

Typical radio resources in wireless networks include frequency channels, time slots, transmission power, etc. Depending on the air interfaces and transmission technologies used in a network, the emphasis for RRM can be different.

The first generation wireless cellular networks transmit analog signals and use FDMA as the multiple access technology. Allocating the frequency channels to different radio cells is the main task for RRM. Each cell is assigned a number of frequency channels from the total available channels, and two cells can reuse the same channels if the distance between them is sufficiently long. Given the total number of available channels, if the frequency reuse distance is shorter (i.e., cells reusing the same frequency channel are closer to each other), more channels can be allocated to each cell, and higher capacity can be achieved in each cell. On the other hand, a shorter frequency reuse distance results in stronger co-channel interference and reduces the received signal quality. Because of these contradictory effects, frequency reuse distance is one of the most important design parameters in the first generation cellular networks. By reducing the cell size, the required transmission power for each user is decreased, which reduces the mutual interference and allows better frequency reuse. The price, of course, is that more BSs should be deployed to cover the same network service area. Typically, the coverage radius for a macrocell is from 1 km to 2 km, for a micro-cell is from 400 m to 2 km, and for a pico-cell is from 4 m to 200 m. Within each cell, resource allocations are rather simple. Each user is allocated a frequency channel, which is dedicated to it throughout its communications.

The second generation wireless cellular networks use digital transmissions, which allow signals from multiple users to share the same frequency channel on TDMA basis. In Global System for Mobile communications (GSM), the channel time is divided into equal length superframes, each of which is further divided into time slots. Each user's connection is allocated a fixed number of time slots. The digital signals together with TDMA-based transmissions allow more flexible resource allocations than in the first generation networks. More time slots can be allocated to data users requiring higher transmission rates. Once a certain number of time slots are allocated to a given user, they cannot be used by other users until the current user completes its communications. This relative static feature of resource allocations is similar to that in the first generation networks.

The second generation cellular networks also include narrow-band CDMA cellular networks, where all the radio cells can reuse the same spectrum, and users in the same and different radio cells can transmit simultaneously. Different users are distinguished by different code channels. As the code channels may not be orthogonal, simultaneous transmissions (both from the same and different cells) can interfere with each other. In such networks, transmission power and interference management becomes more important, and the main purpose of power control is to ensure all users to receive satisfactory signal-to-interference-and-noise ratio (SINR). As the activity of voice and data traffic is intermittent, the transmission power or rate of users can be dynamically adjusted based on current interference conditions.

The third generation cellular networks are CDMA-based. Comparing to the CDMA-based second generation cellular networks, the third generation networks use wider bandwidth to achieve higher transmission rates. Transmission power control is still one of the most important aspects for resource allocations, but is often combined with time slot allocations, allowing more dynamic and efficient resource utilization. This makes the networks suitable to support traffic with different characteristics and QoS requirements. Complicated traffic scheduling schemes are necessary to both satisfy the QoS requirements of the users and efficiently utilize the network resources. The transmission time and power of individual users can be changed dynamically based on both short and long term requirements of the users and other network conditions.

With the increasing popularity of wireless networks, the demands for wireless communication services also increases. This pushes the emergence of new air interfaces, transmission techniques, and access technologies. The fourth generation wireless communication networks are expected to integrate these heterogeneous networks to support a wide range of applications. For example, by interconnecting wireless local area networks (WLANs), wireless cellular networks, and Worldwide Interoperability for Microwave Access (WiMAX) networks, the fourth generation wireless networks can serve users with different transmission ranges, throughput requirements, and mobility patterns. The user's terminals may even automatically switch among these air interfaces based on certain criteria, such as QoS, cost, or other agreements with the network providers. Resources can be shared among different air interfaces to optimize certain objectives, such as overall revenue of the network providers or users' satisfaction, or a combination of the two.

1.1.3 Traffic QoS and Scheduling

The resource allocations in the first and second generation wireless networks are static, simple and suitable for supporting the wireless communication traffic in the early stage, when the services are mainly voice conversations and low rate data. With the increasing popularity of wireless mobile communications, wireless communication networks were demanded to serve a wide range of applications, such as video conferencing and high speed data transmissions. In a lot of cases, a connection may have active traffic for only very short time intervals, and does not generate any traffic during other time intervals. For some traffic, the peak transmission rate can be significantly higher than the average rate. As a result, the fixed resource allocations based on the peak rate can lead to very low resource utilization, since the allocated resource is wasted when the users are transmitting at lower rates or not having active traffic at all. On the other hand, allocating the radio resources based on the average transmission rate results in poor QoS, which can be either significantly long transmission delay or very high packet loss rate, none is acceptable to the users.

In order to efficiently utilize the available radio resources for supporting more users with different types of traffic, the network resources should be shared by the users more dynamically. When one user is not having active traffic, the channel resources can be allocated to other users. For users requiring variable service rates, the amount of allocated resources should be changed based on the current requirements. While such dynamic resource management provides opportunities for better resource utilization, it requires more complicated strategies for effectively protecting the QoS to the users. This is especially important for the traffic requiring strict QoS. When serving this type of traffic, the objective for resource allocations is mainly to guarantee the required QoS using a minimum amount of radio resources, so that as many users as possible can be served in the network. In contrast, some traffic may only require best effort services, and the objective of RRM for serving this type of traffic can be defined to maximize the network resource utilization, subject to a certain level of fairness among the QoS provided to the users. The exact QoS requirements of a user's traffic depends on specific applications, and may also vary depending on the subjective feelings of the user.

Table 1.1 shows some example QoS parameters for several applications. Based on the latency requirements, the user's traffic can be divided into real-time and non-real-time. Real-time traffic has stringent latency requirements and requires timely delivery. Typical real-time traffic includes voice and video traffic. Voice traffic includes alternate talk and silent spurts, and usually has constant packet generation rate during the talk spurts. Real-time voice conversation requires both short transmission delay and delay jittet. For video traffic, the source generated traffic is usually compressed, and the compression rate may change from frame to frame. As a result, video traffic often has varying pack sizes and highly variable packet rates. Interactive video traffic, such as videoconferencing, requires both short delay and small delay jitter. Streaming video can tolerate much longer delay, and usually has no significant jitter requirements. Although both voice and video traffic can tolerate some packet losses (either due to buffer overflow or transmission errors), most data traffic requires very low packet loss rate. Data traffic can usually tolerate much longer delay than real-time traffic. For example, a few seconds of transmission delay can hardly be noticed for Internet downloading. Some data traffic may require a minimum throughput over a period of time, and other data traffic only needs best effort services.

Because of the strict latency requirement, real-time traffic has to be delivered before it expires. This can consume a lot of network resources when the transmission

Table 1.1 Typical QoS requirements for different applications

Applications	Loss rate	Delay	Jitter
VoIP	$\leq 1\%$	≤ 150ms	≤ 30ms
Interactive video	$\leq 1\%$	≤ 150ms	≤ 30ms
Streaming video	$\leq 5\%$	4-5s	no requirement
Web browsing	very low	2 s	no requirement
Internet Download	very low	200 s	no requirement

condition is poor, which reduces the resource utilization. On the other hand, data traffic can be buffered and wait for good transmission opportunities, e.g., good link gains or low interference level. The work to decide which users should transmit at what time and be allocated to how much network resources is called traffic scheduling. The scheduling decisions are usually updated periodically, depending on how fast the network conditions change, including physical channel fading, traffic volume, user locations, etc. Such traffic scheduling can be performed in centralized or distributed ways. In a centralized scheduling, a central station collects all necessary information, makes transmission decisions, and then informs the transmitters of their transmission time and power. The central station can be the BS in a cellular network or an access point (AP) in a WLAN. Information required for making the scheduling decisions can include link gains, interference conditions, amount of backlogged traffic, QoS requirements, priority levels of the users, and other related information. Centralized scheduling cannot be performed in networks where there is no central station available, such as ad hoc networks. In some networks, a central station is available, but the overhead can be high for collecting necessary information to make accurate central scheduling decisions. In these networks, individual nodes may need to measure or estimate the channel and interference conditions locally, exchange information with their neighbors, and make transmission decisions based on limited information available to them.

1.1.4 Connection Admission Control

A user's connection usually lasts for much longer time than a single packet transmission time. For QoS traffic, the network should guarantee the transmission quality during the entire lifetime of the connections. In order to achieve this objective, the amount of QoS traffic in a network must be controlled to be below the network capacity, not only for a given time slot, but also throughout the lifetime of the admitted connections. The function of admission control is to limit the amount of the traffic over a long term by making admission decisions at the time of a connection request. Before establishing a connection, a user should send a connection request to the network, specifying the traffic characteristics and the required QoS, and the network then decides whether such a request can be satisfied, based on the amount of available resources in the network. The traffic characteristics may include average and peak packet generation rates, burstiness, and other parameters, and the QoS requirement may include packet loss rate, average or maximum delay, and delay jitter.

In most cases, the duration of a user's connection is random and unknown at the time the user makes the connection request. The requirement for admission control is that once a connection is admitted into the system, its QoS can be guaranteed throughout its lifetime. Based on this, there are two basic criteria for admission control. First, a connection request can be accepted into the system only if there is a sufficient amount of resources in the system to satisfy its required QoS. Second, all existing connections should still be able to receive their required QoS after admission of the new connection. In a network where the radio resources allocated to different users are orthogonal, the first criterion implies the second one. For example, in an FDMA or TDMA-based network without frequency reuse, as long as there is a sufficient number of frequency channels or time slots for a new connection, the new connection can be admitted without affecting the QoS of any existing connections. On the other hand, this may not be true in other networks. For example, when the frequency channels can be reused, transmissions from different users can interfere with each other. In this case, admitting a new connection increases the interference level and reduces the QoS of the existing users. QoS of the existing users cannot be maintained unless they all raise their transmission power, which may not be possible if they have already reached the maximum transmission power limit. Similar problems also exist in networks where users can share the time slots, in which case admitting new users may increase data transmission delay of existing users.

Effectiveness and efficiency are two metrics to evaluate the performance of radio resource management. Effectiveness is to guarantee QoS of the admitted traffic, and efficiency is to maximize the amount of traffic admitted into the system. In a practical system, the amount of resources required by a certain user also depends on specific scheduling schemes used, i.e., how radio resources are allocated among the users. For example, the amount of required resources for a variable bit rate connection can be significantly different, depending on whether the resource is allocated based on its peak rate or instantaneous rate. Most practical scheduling schemes are complicated, and include a lot of details in order to consider different network and traffic conditions. It is very difficult, if not impossible, to both maximize the number of admitted QoS connections and guarantee their required QoS all the time [4]. A practical admission control scheme may either over-estimate or under-estimate the amount of traffic that can be served in the network. In the former case, it depends on the scheduling process to decline the resource usage for some users; and in the latter case, the extra network resource may still be utilized by serving best effort traffic [5]. There is a tradeoff between the connection level capacity and the packet level throughput. For an extreme case, if the amount of admitted QoS traffic is maximized, the system may have no resource to serve any best effort traffic.

1.2 Power Control and Interference Management

1.2.1 Power Control and Its Relationship with Other RRM Functions

Transmission power is one type of the basic radio resources in wireless communication networks. Sufficient transmission power is required in order to keep the signal strength at the receiver above a certain level, so that the received signal can be recovered with an acceptable bit error rate (BER). Power control is a function that controls the power levels at the transmitter or receiver ends in order to achieve certain QoS requirements of individual users. In the literature, power control can often be exchanged with two other terminologies, power distribution and power allocations, especially when the emphasis is regarding the target power levels. In some other cases, the use of power control emphasizes more on *how* the target power levels are achieved — for example, through a centralized controller or using distributed iterations.

When multiple users share the same frequency channel, simultaneous transmissions interfere with each other. In order to keep the BER below a certain level, a minimum SINR should be kept at the receiver input. Transmission power of the users should be controlled in order to both protect the receiving quality of their own transmissions and limit the interference to other users. Therefore, power control in such a scenario is also interference control. Any network condition variations (for example, random channel fading) that affect the transmission power of one link may change the interference level to other links, which changes the minimum required signal strength at the receivers of these links, and further changes the required transmission power of these links. For example, a node may increase its transmission power to combat reduced channel gain or to achieve better signal quality at the receiver end. At the same time, this increases the interference to other links sharing the same channel and reduces their receiving quality. As a result, all other transmitters may increase their transmission power. As the number of users increases, co-channel interference increases, which requires each node to transmit higher power. If the total traffic load is within the network capacity, this type of mutual effect may eventually reach a balance, and QoS of all the links can be satisfied. Otherwise, some users may have to stop their transmissions.

Transmission power determines the maximum transmission range of a node. In a cellular network, nodes close to the cell boundary should transmit higher power in order to reach the BS. The maximum transmission power of the nodes determines not only the cell coverage, but also the frequency reuse distance. The former affects the BS deployment, and the latter is directly related to the system capacity. For example, pico-cellular networks can achieve much higher capacity than micro- and macro-cellular networks, but at a price of more densely deployed BSs. In multi-hop networks, higher transmission power can result in a smaller number of hops from the source to the destination, and lower transmission power increases the spatial multiplexing and may improve the system capacity. In a wireless mesh network, controlling the transmission power can also affect network connectivity and routing.

Transmission power is directly related to energy consumption of the nodes. Reducing transmission power not only reduces interference, but also decreases the energy consumption of the battery, given the same amount of transmission time. This issue is more important to the handsets than to the BSs in traditional cellular communication networks, where the BSs are AC powered and their energy consumption is not much concerned. With the popularity of wireless communications, wireless communication networks are often deployed in places where AC power supplies are unavailable, and both the infrastructure and the end equipment have to be powered by batteries or solar/wind powered batteries [6, 7]. Furthermore, in some networks, the replacement or recharge of the batteries may be inconvenient. In such cases, transmission power control is important not only to the battery lifetime, but also to the network lifetime.

1.2.2 Feasibility Analysis for Power Control

Consider a network having N active links at a given time, all sharing the same frequency channel. We refer the transmitter and receiver of the ith link as the ith transmitter and ith receiver, respectively. Denote the link gain from the ith transmitter to the jth receiver as g_{ij}. Let P_i denote the transmission power of the ith transmitter, then the SINR at the ith receiver is given by

$$\gamma_i = \frac{P_i g_{ii}}{\sum_{j=1, j\neq i}^{N} P_j g_{ji} + \eta_i}, \tag{1.1}$$

where η_i is the background noise power at the ith receiver. Let γ_i^* be the minimum required SINR for link i. We have

$$\frac{P_i g_{ii}}{\sum_{j=1, j\neq i}^{N} P_j g_{ji} + \eta_i} \geq \gamma_i^*. \tag{1.2}$$

From (1.2) we have

$$P_i \geq \sum_{j=1, j\neq i}^{N} \gamma_i^* \frac{g_{ji}}{g_{ii}} P_j + \gamma_i^* \frac{\eta_i}{g_{ii}} \tag{1.3}$$

for all $i = 1, 2, \ldots, N$. By defining a set of vectors and matrices, the N expressions defined by (1.3) can be rewritten in a matrix form. We first define column vectors $\mathbf{P} = (P_1, P_2, \ldots, P_N)^T$ and $\mathbf{B} = (B_1, B_2, \ldots, B_N)^T$ with B_i given by

$$B_i = \gamma_i^* \frac{\eta_i}{g_{ii}}. \tag{1.4}$$

We then define three $N \times N$ matrices, an identity matrix $\mathbf{I}$, a diagonal matrix $\boldsymbol{\Gamma} = \text{diag}(\gamma_1^*, \gamma_2^*, \ldots, \gamma_N^*)$, and matrix $\mathbf{G} = (G_{ij})$ with the ith-row and jth-column element given by

$$G_{ij} = \begin{cases} \frac{g_{ji}}{g_{ii}}, & \text{when } 1 \le i, j \le N \text{ and } i \ne j \\ 0, & \text{when } 1 \le i = j \le N, \end{cases} \tag{1.5}$$

where G_{ij} is referred to as normalized link gain from link j to link i. Then (1.3) can be rewritten as

$$(\mathbf{I} - \boldsymbol{\Gamma}\mathbf{G})\mathbf{P} \succeq \mathbf{B}, \tag{1.6}$$

where $\succeq$ is element-wise larger than or equal to.

Existence of solutions

When the inverse of $(\mathbf{I} - \boldsymbol{\Gamma}\mathbf{G})$ exists, $\mathbf{P}$ can be solved from (1.6) as

$$\mathbf{P} \succeq (\mathbf{I} - \boldsymbol{\Gamma}\mathbf{G})^{-1} \mathbf{B} = \sum_{k=0}^{\infty} (\boldsymbol{\Gamma}\mathbf{G})^k \mathbf{B}. \tag{1.7}$$

Since all elements in $\boldsymbol{\Gamma}$ and $\mathbf{G}$ are non-negative, all the elements in the product $\boldsymbol{\Gamma}\mathbf{G}$ are also non-negative. Furthermore, all elements in $\mathbf{B}$ are non-negative. Because of this, the power vector is also non-negative. In this case, we say the power control problem has a feasible solution, or the target SINR $\boldsymbol{\Gamma}$ is achievable.

Note that the condition required for the existence of $(\mathbf{I} - \boldsymbol{\Gamma}\mathbf{G})^{-1}$ is that the Perron-Frobenius eigenvalue (or the largest eigenvalue) of the matrix $\boldsymbol{\Gamma}\mathbf{G}$, denoted by $\rho(\boldsymbol{\Gamma}\mathbf{G})$, is less than or equal to one, which is also the condition for the power control problem to be feasible.

When $\rho(\boldsymbol{\Gamma}\mathbf{G}) = 1$, $|\mathbf{I} - \boldsymbol{\Gamma}\mathbf{G}| = 0$. The problem has a feasible solution only when $\mathbf{B}$ is a zero vector, i.e., $\eta_i = 0$ for all $i = 1, 2, \ldots, N$.

Minimum transmission power

Consider $\mathbf{P}^* = (P_i^*, P_2^*, \ldots, P_N^*)^T$. When $\mathbf{P} = \mathbf{P}^*$, the equality in (1.6) holds. That is

$$(\mathbf{I} - \boldsymbol{\Gamma}\mathbf{G})\mathbf{P}^* = \mathbf{B}. \tag{1.8}$$

Then for all $i = 1, 2, \ldots, N$, P_i^* is the minimum power that the ith transmitter should transmit.

In order to prove this, consider another vector $\mathbf{P}_1$ that satisfies (1.6), and $\mathbf{P}_1 \ne \mathbf{P}^*$. Then

$$(\mathbf{I} - \boldsymbol{\Gamma}\mathbf{G})(\mathbf{P}_1 - \mathbf{P}^*) = \mathbf{B}_1, \tag{1.9}$$

where $\mathbf{B}_1$ is a non-negative vector and has at least one element larger than zero. From (1.9) we have

$$\mathbf{P}_1 - \mathbf{P}^* = (\mathbf{I} - \boldsymbol{\Gamma}\mathbf{G})^{-1} \mathbf{B}_1 = \sum_{k=0}^{\infty} \boldsymbol{\Gamma}\mathbf{G}^k \mathbf{B}_1 \succ \mathbf{0}, \tag{1.10}$$

where $\succ$ is element-wise larger than.

Maximum target SINR for homogeneous traffic

When there is no feasible solution, i.e., at least one element of $\mathbf{P}$ is less than 0, the target SINRs cannot be supported. In this case, at least one user should reduce its target SINR or be removed (stop its transmissions). For homogeneous traffic, $\gamma_i^* = \gamma^*$ for all $i = 1, 2, \ldots, N$, and (1.6) becomes

$$(\mathbf{I} - \gamma^* \mathbf{G}) \mathbf{P} \succeq \mathbf{B}, \tag{1.11}$$

which has a feasible solution if and only if $\rho(\gamma^* \mathbf{G}) \leq 1$, or

$$\gamma^* \leq \frac{1}{\rho(\mathbf{G})}. \tag{1.12}$$

That is, $1/\rho(\mathbf{G})$ is the maximum achievable SINR. When $\eta_i = 0$ for all $i = 1, 2, \ldots, N$, $\mathbf{B} = \mathbf{0}$, and from (1.11) we have

$$\frac{1}{\gamma^*} \mathbf{P} \succeq \mathbf{GP}, \tag{1.13}$$

from which we can see that in an interference limited system (neglecting η_i's), the optimum power vector equals the Perron-Frobenius eigenvector of $\mathbf{G}$.

Link removal

When there is no feasible solution to achieve the target SINRs, at least one link should be removed, and the removed link is in communication outage. Different criteria can be used for deciding which links to be removed so that a feasible solution exists for the remaining links to achieve the target SINRs.

When there are at least two links in a system, it can be observed that matrix $\boldsymbol{\Gamma}\mathbf{G}$ is non-negative (element-wise) and irreducible, since the diagonal elements are all zero and all other elements are greater than zero. For such matrices, the Perron-Frobenius theorem [8] indicates that $\rho(\boldsymbol{\Gamma}\mathbf{G})$ may increase when any elements of $\boldsymbol{\Gamma}\mathbf{G}$ increase. (See Appendix A for some descriptions about the Perron-Frobenius theorem.) Therefore, larger elements of $\boldsymbol{\Gamma}\mathbf{G}$ lead to a higher possibility that $\rho(\boldsymbol{\Gamma}\mathbf{G}) > 1$, and a higher chance that no feasible solution exists for $\mathbf{P}$. Given the target SINRs, large elements in $\boldsymbol{\Gamma}\mathbf{G}$ may be due to i) large γ_i^*, ii) poor link gain between the desired transmitter and receiver of a link (i.e., small g_{ii}), or iii) strong interference between links (large g_{ij} for $i \neq j$). Based on this, link removal criteria can be designed according to different objectives, such as minimizing the overall outage probability or keeping fair outage probabilities among different links, so that more

links can transmit for the given objective. This will be discussed more in Chapters
2 and 4.

Iterative Power Control

We have discussed how to solve the transmission power vector mathematically.
In a practical system, there should be an approach to guiding individual transmitters
to reach the target power levels. This may be implemented in either centralized or
distributed ways. In a centralized implementation, a central station, such as the BS in
a cellular network, can communicate with all transmitters in the network, collect the
link gain information, find the normalized link gain matrix $\mathbf{G}$ and the interference
vector $\mathbf{B}$, calculate the transmission power, and inform the transmitters of the calculated target power levels. When this type of central stations are unavailable, such as
in ad hoc networks, distributed power control may be implemented, usually through
iterations. The dynamic and distributed power control scheme proposed in [9] is
briefly introduced here.

The objective of the iterative power control is for each node to transmit the minimum power, i.e., the equality holds in (1.6). That is,

$$(\mathbf{I} - \boldsymbol{\Gamma}\mathbf{G})\mathbf{P} = \mathbf{B}, \qquad (1.14)$$

which can be further rewritten as

$$\mathbf{P} = \boldsymbol{\Gamma}\mathbf{G}\mathbf{P} + \mathbf{B}. \qquad (1.15)$$

Let $\mathbf{P}(k)$ be the vector of the actual transmission power in the kth iteration. An
iterative formula can be designed based on (1.15) to find the transmission power
vector in the next iteration as

$$\mathbf{P}(k+1) = \boldsymbol{\Gamma}\mathbf{G}\mathbf{P}(k) + \mathbf{B}, \qquad (1.16)$$

from which the transmission power of transmitter i is given by

$$P_i(k+1) = \frac{\gamma_i^*}{g_{ii}} \left[\sum_{j=1, j\neq i}^{N} g_{ji}P_j(k) + \eta_i \right] \qquad (1.17)$$

$$= \gamma_i^* \frac{\sum_{j=1, j\neq i}^{N} g_{ji}P_j(k) + \eta_i}{g_{ii}P_i(k)} P_i(k) \qquad (1.18)$$

$$= \frac{\gamma_i^*}{\gamma_i(k)} P_i(k), \qquad (1.19)$$

where $P_i(k)$ is the transmission power of transmitter i in the kth iteration, and

$$\gamma_i(k) = \frac{g_{ii}P_i(k)}{\sum_{j=1, j\neq i}^{N} g_{ji}P_j(k) + \eta_i} \qquad (1.20)$$

is the actual SINR at the ith receiver and can be measured at the receiver. From (1.19) we can see that the transmission power in the next iteration is the transmission power in the current iteration multiplied by the ratio of the target SINR to the actual SINR in the current iteration.

To implement this iterative power control in a practical network, a feedback channel is required from the receiver to the transmitter. The receiver measures $\gamma_i(k)$, and then passes the value to its transmitter through the feedback channel. Unlike in a centralized power control, link gains from both the desired and interfering transmitters are not required during this power control process, which is an attractive feature that makes the dynamic power control method relatively easy to implement. Like any iterative algorithms, convergence is an important issue that should be considered in order to evaluate performance of the algorithm. The convergence of the above power control algorithm is proved in [9]. When the maximum transmission power is limited, the actual transmission power in each iteration takes either the value from (1.19) or the maximum power, whichever is smaller. This is the distributed constrained power control algorithm in [10].

The above dynamic power control does not require a central controller, and can be implemented distributively by individual links. An important condition for applying this dynamic power control is that each link should have a target SINR, i.e., γ_i^*. This makes it difficult to be applied in some network scenarios. The first scenario is multihop networks, where the target SINRs of the multiple hops along an end-to-end path of a given connection are often correlated in order to achieve a certain QoS at the destination. Another scenario is when multiple transmitters send the same message to the same receiver, either at the same or different time, and the receiver should recover the original message after combining the multiple copies of the received signals. In this case, the multiple transmissions jointly affect the SINR at the receiver. In both the cases, it is unlikely that a simple power control algorithm can optimize the transmission power of all the transmitters, including the source and the intermediate relay nodes in the first case and the multiple transmitters in the second case. More complicated power control schemes are required for these cases.

1.2.3 More on SINR Achievability

At the end of this section we consider a network with N existing links, and a new link requests to join the network. Given the target SINRs of the existing links, what is the maximum achievable SINR for the new link? Knowing the answer can help make admission control decisions for the new connection request. In [11] this problem is studied in a two-tier scenario, where N femto-cells and one macro-cell coexist in the same geographical area. Below we extend part of the analysis to a more general scenario. Using the same notations as in the previous subsection we define $\Gamma = \mathrm{diag}(\gamma_1^*, \gamma_2^*, \ldots, \gamma_N^*)$ as the SINR matrix, and $\mathbf{G} = (G_{ij})$ as the normalized link gain matrix for the existing links. Denote the new link as link 0. Assuming the new link is already in the system, we define $\Gamma' = \mathrm{diag}(\gamma_0^*, \gamma_1^*, \gamma_2^*, \ldots, \gamma_N^*)$, and $\mathbf{G}' = (G'_{ij})$ with the ith how and jth row given by

$$G'_{ij} = \begin{cases} \frac{g_{ji}}{g_{ii}}, & \text{when } 0 \leq i,j \leq N, i \neq j \\ 0, & \text{when } 0 \leq i = j \leq N. \end{cases} \tag{1.21}$$

We can write the product of the two matrices $\boldsymbol{\Gamma}'\mathbf{G}'$ as

$$\boldsymbol{\Gamma}'\mathbf{G}' = \begin{pmatrix} 0 & \gamma_0 \mathbf{G}_0^T \\ \boldsymbol{\Gamma}\mathbf{G}_e & \boldsymbol{\Gamma}\mathbf{G} \end{pmatrix}, \tag{1.22}$$

where the vector $\mathbf{G}_0^T = (G'_{01}, G'_{02}, \ldots, G'_{0N})$ consists of the normalized link gains from the existing links to the new link, and $\mathbf{G}_e = (G'_{10}, G'_{20}, \ldots, G'_{N0})^T$ consists of the normalized link gains from the new link to the existing links.

Denote $\chi = \rho(\boldsymbol{\Gamma}'\mathbf{G}')$. It is obvious that $\rho(\boldsymbol{\Gamma}\mathbf{G}) \leq \chi$. That is, adding the new link will drive the network towards being infeasible.

Given that the SINR's are achievable before the new link is admitted in the system, i.e., a feasible power solution exists for the existing N links, the following conclusions have been proved in [11].

- The highest SINR that the new link can achieve is given by

$$\gamma_0 = \frac{\chi^2}{\mathbf{G}_0^T[\mathbf{I} - (\boldsymbol{\Gamma}/\chi)\mathbf{G}]^{-1}\boldsymbol{\Gamma}\mathbf{G}_e}. \tag{1.23}$$

- Given the target SINR of the new link as γ_0^*, a necessary condition for having a feasible power solution after adding the new link is given by

$$\gamma_0^* \leq \frac{1}{\mathbf{G}_0^T \boldsymbol{\Gamma}\mathbf{G}_e}. \tag{1.24}$$

- Assuming all the existing links have the same target SINR γ^* and $\gamma^* < 1/\rho(\mathbf{G})$, the following relationship holds,

$$\gamma_0^*\gamma^* \leq \frac{1}{\mathbf{G}_0^T\mathbf{G}_e}. \tag{1.25}$$

As an example, consider there is only one existing link (link 1), then the matrix $\boldsymbol{\Gamma}'\mathbf{G}'$ is given by

$$\boldsymbol{\Gamma}'\mathbf{G}' = \begin{pmatrix} 0 & \gamma_0^* G'_{01} \\ \gamma_1^* G'_{10} & 0. \end{pmatrix} \tag{1.26}$$

We can find $\rho(\boldsymbol{\Gamma}'\mathbf{G}') = \sqrt{\gamma_0^*\gamma_1^* G'_{01}G'_{10}}$. In order for the SINRs to be achievable,

$$\gamma_0^*\gamma_1^* < \frac{1}{G'_{10}G'_{01}}. \tag{1.27}$$

That is, the product of the target SINRs of the existing and new links is limited by the inverse product of the cross-link gains.

References

1. Grossglauser M, Tse DNC (2002) Mobility increases the capacity of ad hoc wireless networks. IEEE/ACM Transactions on Networking 10(4): 477 - 486.
2. Diggavi SN, Grossglauser M, Tse DNC (2005) Even one-dimensional mobility increases the capacity of wireless networks. IEEE Transactions on Information Theory 51(11): 3947 - 3954.
3. Cheng HT, Zhuang W (2009) QoS-driven MAC-layer resource allocation for wireless mesh networks with non-altruistic node cooperation and service differentiation. IEEE Transactions on Wireless Communications 8(12): 1-14.
4. Abdrabou A, Zhuang W (2008) Stochastic delay guarantees and statistical call admission control for IEEE 802.11 single-hop ad hoc networks. IEEE Transactions on Wireless Communications 7(10): 3972-3981.
5. Leong CW, Zhuang W, Cheng Y, Wang L (2006) Optimal resource allocation and adaptive call admission control for voice/data integrated cellular networks. IEEE Transactions on Vehicular Technology 55(2): 654-669.
6. Zhang F, Todd TD, Zhao D, Kezys V (2006) Power saving access points for IEEE 802.11 wireless network infrastructure. IEEE Transactions on Mobile Computing 5(2): 144-156.
7. Cai LX, Liu Y, Luan H, Shen X, Mark JW, Poor HV, Dimensioning network deployment and resource management in green mesh networks, IEEE Wireless Communications, to appear.
8. Varga RS (1962) Matrix iterative analysis, Chapter 2, Prentice-Hall, Inc., Englewood Cliffs, N.J.
9. Grandhi SA, Vijaya R, Goodman DJ (1994) Distributed power control in cellular radio systems. IEEE Transactions on Communications 42(5): 226-228.
10. Grandhi SA, Zander J, Yates RD (1995) Constrained power control. Wireless Personal Communications 1(4): 257-270.
11. Chandrasekhar V, Andrews JG, Muharemovic T, Shen Z, Gatherer A (2009) Power control in two-tier femtocell networks. IEEE Transactions on Wireless Communications 8(8): 4316-4328.

Chapter 2
CDMA-Based Wireless Cellular Networks

Abstract In a CDMA-based cellular network, all radio cells share the same frequency bands, and users can transmit simultaneously. Transmissions from one user causes interference to other users. The more users are in the system and the higher power they transmit, the more interference they generate to one another. A CDMA-based system is typically interference-limited. The management of transmission power and mutual interference is directly related to system capacity and quality-of-service (QoS) to the users. In this chapter, we first briefly review the motivations of power control in cellular CDMA networks, then study the power allocation problem in a single-cell CDMA network. Based on the analysis, different aspects that affect the transmission power and system capacity are investigated. We then study the power allocation problem in a multi-cell CDMA network. The relationship between transmission power, network capacity, and QoS to the users is analyzed.

Keywords: CDMA, cellular network, power control, SINR, pole capacity, outage, soft handoff.

2.1 Motivations for Power Control

In a CDMA-based network, each transmitter uses a unique spreading code to generate the transmitted signals. The intended receiver can reproduce the spreading code used by the transmitter and recover the desired signals. The cross-correlation of different spreading codes is ideally zero, so that the desired signal can be recovered and interfering signals can be removed at the receiver. In a practical system, the radio channel can be non-linear, and the spreading codes may not be orthogonal to one another. Therefore, transmissions of the users can cause interference to one another. The signal-to-interference-plus-noise ratio (SINR), defined as the desired signal power divided by the total power of interference and noise, can be used to evaluate the received signal quality.

Power control is originally used to solve the near-far problems in the uplink of cellular CDMA networks, where homogeneous traffic (mainly voice) is supported.

17

In the uplink, all the transmissions in the same cell share the same receiver, which is the base station (BS). If different users transmit at the same power, their signals arrive at the BS receiver with different strength. On average, signals from the users far away from the BS are weaker than that from the users close to the BS. Therefore, the SINR of the former can be much worse than that of the latter. That is, the signals from the users near the BS can block the transmissions of the signals from the users far away from the BS. This is the "near-far" effect. In order to balance the received SINRs for the signals from different users, power control is applied. The main purpose of power control is for the signals from different users to arrive at the BS with the same and acceptable SINR. The transmission power of each user is controlled and adjusted based on its channel gain to the BS. Users with poorer channel gain to the BS should transmit higher power.

The function of power control in the downlink is different. Since the signal and interference from the BS arrive at a given mobile station (MS) go through the same radio channel and undergo the same attenuation, power control is not needed to combat the near-far problem. Instead, it is used to provide more power to users located near the cell borders, where the users can suffer from high interference from the transmissions in nearby cells. In addition, the transmission power of the BS should be controlled in order to reduce the interference to nearby cells.

Consider N users transmitting to the same BS. Their signals are all power controlled to have the same power S at the BS receiver input. Each user's signal experiences interference from the transmissions of all the other $N - 1$ users, the total interference power at the receiver is $I = (N - 1)S$, and the SINR of the signal is $S/(I + \eta)$, where η is the background noise power. Let R be the information bit rate, and W be the spreading bandwidth. Then $E_b = S/R$ gives the energy per information bit, and $I_0 = (I + \eta)/W$ gives the interference-plus-noise power spectrum density. The ratio of energy per bit to interference-plus-noise power spectral density (E_b/I_0) is the SINR normalized to each transmission bit, and is commonly used for evaluating the receiving quality of the CDMA users. This is based on a fairly reasonable assumption that the bit-error-rate (BER) at a receiver is a monotonically decreasing function of E_b/I_0. When the noise power is zero, the expression for E_b/I_0 is given by

$$E_b/I_0 = \frac{S/R}{(N-1)S/W} = \frac{W/R}{N-1}, \tag{2.1}$$

where the quantity W/R is called the processing gain, which is a basic parameter for spread spectrum communications. If γ^* is the minimum required value for E_b/I_0 (corresponding to a maximum acceptable BER for a given physical layer design), then N can be solved as

$$N \leq 1 + \frac{W}{R\gamma^*} \triangleq N_p, \tag{2.2}$$

where N_p is referred to as the single cell pole capacity. From (2.2) we can find that for given spreading bandwidth W, a higher pole capacity is achieved if users require lower transmission rate and target γ^*. The pole capacity provides an important upper bound for the capacity, and is independent of the channel conditions of individual links. It is also the maximum capacity for each cell in a multi-cell network. The pole capacity can never be achieved in a real network, since it is obtained by assuming zero noise and interference power, because of which the pole capacity is independent of S. On the other hand, each receiver requires a minimum signal power in order to detect and decode the desired signals. When the link condition is poor, the required transmission power can be high in order to guarantee the minimum power at the receiver end. The actual capacity of the system is then limited by the maximum transmission power of the MSs. In addition, different users may require different transmission rates and target SINRs, which also affect the target receiving and transmission power of each user's signal. This scenario is studied in the next section.

In the remaining part of this book, all SINRs are normalized to per information bit. In another word, they are in fact energy per bit to interference-plus-noise power spectral density ratio.

2.2 Power Allocations in a Single Cell Network

We consider an FDD-based network, where different frequency bands are used for the uplink and the downlink transmissions, and therefore, there is no interference between the uplink and the downlink transmissions. This allows us to study the uplink and the downlink performance separately. We use R_i to denote the transmission rate and γ_i^* to denote the the minimum required SINR for user i, where $i = 1, 2, \ldots, N$, and N is the total number of active users.

Uplink Power Analysis

Each user may have a different rate and SINR requirement, and the target receiving power for different users may be different at the BS. Let $S_{u,i}$ be the target receiving power of user i's signal, and η be the noise power at the BS receiver. The relationship between the target receiving power and the required SINR is given by

$$\frac{W}{R_i} \frac{S_{u,i}}{\sum_{j=1, j \neq i}^{N} S_{u,j} + \eta} \geq \gamma_i^*, \tag{2.3}$$

for all $i = 1, 2, \ldots, N$. This formulation is equivalent to (1.2), if the processing gain (W/R_i) is equal to 1 in (2.3) and all the background noise powers are the same in (1.2). Based on the analysis for (1.2), we know that the minimum power is required for each user when equality holds in (2.3), i.e.,

$$\frac{W}{R_i} \frac{S_{u,i}}{\sum_{j=1, j \neq i}^{N} S_{u,j} + \eta} = \gamma_i^*. \tag{2.4}$$

With some simple manipulations, (2.4) can be rewritten as

$$\sum_{j=1}^{N} S_{u,j} + \eta = \left(\frac{W}{\gamma_i^* R_i} + 1\right) S_{u,i} \tag{2.5}$$

for all $i = 1, 2, \ldots, N$. Define $Q_i = \frac{W}{\gamma_i^* R_i} + 1$, (2.5) becomes

$$\sum_{j=1}^{N} S_{u,j} + \eta = Q_i S_{u,i}. \tag{2.6}$$

Since the left-hand side of (2.6) does not depend on individual user's parameters, the right-hand side of the equation should be the same for different users. That is,

$$Q_i S_{u,i} = Q_j S_{u,j} \tag{2.7}$$

for all $i, j = 1, 2, \ldots, N$. Then $S_{u,j}$ can be written as

$$S_{u,j} = S_{u,i} \frac{Q_i}{Q_j}. \tag{2.8}$$

Replacing $S_{u,j}$ in (2.6) with the right-hand side in (2.8), we can solve $S_{u,i}$ as

$$S_{u,i} = \frac{\eta}{Q_i \left(1 - \sum_{j=1}^{N} \frac{1}{Q_j}\right)}. \tag{2.9}$$

Dividing $S_{u,i}$ by $g_{u,i}$, which is the link gain from MS i to the BS, we can find the required transmission power from user i as

$$P_{u,i} = \frac{\eta}{Q_i \left(1 - \sum_{j=1}^{N} \frac{1}{Q_j}\right) g_{u,i}}. \tag{2.10}$$

From the above derivations we have the following observations:

- Without considering the maximum transmission power limit, a feasible solution to the power allocation problem exists (i.e., all $S_{u,i}$'s are non-negative) if and only if

$$\sum_{j=1}^{N} Q_j^{-1} < 1. \tag{2.11}$$

  If the total capacity of the system is normalized to 1, which is the right-hand side of (2.11), Q_j^{-1} can be considered as the normalized amount of resources consumed by user j. Because of this property, (2.11) can be used as a criterion for admission control in cellular CDMA networks.
- For homogeneous traffic, all the users require the same rate and SINR, $Q_i = N_p = W/(\gamma^* R) + 1$ for all $i = 1, 2, \ldots, N$, is the pole capacity derived in the previous section. In order to have a feasible solution for the receiving power, $N/N_p < 1$

should always hold, or $N < N_p$. The required target receiving power is then given by

$$S_{u,i} = \frac{\eta}{N_p - N},$$ (2.12)

which indicates that the required power increases with the number of users supported in the network.

- The target receiving power for the ith user is inversely proportional to Q_i. This is straightforward, since higher transmission rate and larger SINR requirement (which results in smaller Q_i) requires the support of higher power.
- The target receiving power at the BS for each user is independent of the link gains. The power control manages the target receiving power, which directly affects the mutual interference among the links.
- The required transmission power from a user is inversely proportional to the link gain between itself and the BS. Users with worse link gains should transmit higher power. Furthermore, the required transmission power for each user does not depend on the link gain of any other links. That is, poor link gain of one link, although results in high transmission power from a particular user, does not affect the transmission power of other links in a single cell network. (This is not the case in a multi-cell network, where inter-cell interference exists.)
- By further looking at (2.12) we can find that $S_{u,i}$ can increase significantly with the number of users when the latter is close to the pole capacity. As a result, the required transmission power also increases, and may exceed the maximum transmission power limit of the MS. When the maximum transmission power is relatively small or the link conditions are poor, the actual capacity can be much smaller than the pole capacity.

Downlink Power Analysis

The power distribution in the downlink is similar to that in the uplink, except that orthogonal codes may be used in the downlink for users associated with the same BS in order to reduce co-channel interference within the cell. However, the orthogonality may not be maintained perfectly at the receiver end due to the non-linearity inherent in the radio channel propagation. A variable ξ is used to denote this effect. When $\xi = 0$, different transmissions are kept orthogonal at the receiver end; and when $\xi = 1$, all power for one user contributes to interference to others. In addition, different receivers may experience different noise power. Let $P_{d,i}$ be the required transmission power from the BS to the ith MS, $g_{d,i}$ be the link gain from the BS to the MS, and η_i be the background noise power at the receiver of MS i. The downlink transmission power should satisfy the following relationship,

$$\frac{W}{R_i} \frac{P_{d,i} g_{d,i}}{\sum_{j=1, j \neq i}^{N} \xi P_{d,j} g_{d,i} + \eta_i} \geq \gamma_i^*.$$ (2.13)

When equality holds in (2.13), the transmission power to each user is minimized. Using a similar approach as for the uplink, we can find the minimum value for $P_{d,i}$ as

$$P_{d,i} = \frac{\eta_i}{Q_{\xi,i}\left(1 - \sum_{j=1}^{N} \frac{1}{Q_{\xi,j}}\right)g_{d,i}}, \tag{2.14}$$

where $Q_{\xi,i} = 1 + \frac{W}{\xi \gamma_i^* R_i}$. Similar to the uplink, $1/Q_{\xi,i}$ can be considered as the normalized amount of resource consumed by user i in the downlink, if the total downlink capacity is normalized to 1. The target receiving power at user i is given by

$$S_{d,i} = P_{d,i} g_{d,i} = \frac{\eta_i}{Q_{\xi,i}\left(1 - \sum_{j=1}^{N} \frac{1}{Q_{\xi,j}}\right)}. \tag{2.15}$$

When $\gamma_i^* = \gamma^*$ and $R_i = R$ for all i, $Q_{\xi,i} = 1 + \frac{W}{\xi \gamma^* R} \triangleq N_{\xi,p}$ is the single cell pole capacity in the downlink for homogeneous traffic.

For homogeneous traffic, we can see that the target transmission power and receiving power in the downlink have similar properties as in the uplink. In addition, if the rates and SINRs in both the directions are the same (this may not be true in a practical system), then we have the following observations:

- When $\xi < 1$, the pole capacity in the downlink is higher than that in the uplink, and the normalized resource consumed by each user in the downlink is smaller than that in the uplink; furthermore, if the channel is reciprocal, i.e., $g_{u,i} = g_{d,i}$, the required transmission power in the downlink is lower than that in the uplink.
- When $\xi = 1$, the pole capacity and the normalized resource consumed by each user in the downlink are the same as in the uplink; furthermore, if the channel is reciprocal, the required transmission power in the uplink is exactly the same as that in the downlink.

Outage

The uplink channel is an access channel, where all the transmitters are distributed in different places and share the same receiver; while the downlink channel is a broadcast channel, where all the links share the same transmitter. Let $P_{max,MS}$ and $P_{max,BS}$, respectively, represent the maximum transmission power of an MS and a BS. In the uplink, communication outage occurs if $P_{u,i} > P_{max,MS}$; and in the downlink, outage occurs if $\sum_{i=1}^{N} P_{d,i} > P_{max,BS}$. Normally, $P_{max,MS} < P_{max,BS}$, and the capacity in the uplink is lower than that in the downlink, or outage in the uplink is higher than that in the downlink, if the users require the same rate and SINR in both directions.

For the uplink, when there are one or multiple users having $P_{u,i} > P_{max,MS}$, removing all these users (i.e. putting these users in outage) would make the power allocation problem feasible for the remaining users. However, it may be not necessary to remove all these users. Based on the analysis for the uplink we can find that i) both high traffic load and poor link condition can result in high transmission power from a user, and ii) removing any user can decrease the target receiving power (and therefore reduce the required transmission power) of all the remaining users. When $P_{u,i} > P_{max,MS}$ for one or multiple users, it depends on specific objectives to make decisions regarding how many and which existing users should be removed in order to make the power allocation feasible for the remaining users. The situation is similar for the downlink when $\sum_{i=1}^{N} P_{d,i} > P_{max,BS}$.

We consider several simple criteria for user removal. For the uplink, when there is at least one user having $P_{u,i} > P_{max,MS}$, we consider two criteria. In the first criterion, all the users with $P_{u,i} > P_{max,MS}$ are removed, and after this the transmission power for the remaining users should be all below $P_{max,MS}$. In the second criterion, the user with the largest $P_{u,i}$ is first removed, and the transmission power for the remaining users is recalculated. If $P_{u,i} > P_{max,MS}$ for any of the remaining users, another user is removed based on the same criterion. This process is repeated until $P_{u,i} \leq P_{max,MS}$ for all the remaining users. For the downlink, the user with the largest $P_{u,i}$ is removed first. After this the transmission power for the remaining users is recalculated. If $\sum_{i=1}^{N} P_{d,i} \leq P_{max,BS}$, the removal process ends. Otherwise, another user is removed based on the same criterion. Outage occurs to the removed users. Figs. 2.1 and 2.2 show the outage probabilities for the uplink and downlink, respectively. These results are generated based on a link gain model that includes both path loss and log-normally distributed channel fading, and the channel is reciprocal in the uplink and the downlink, i.e., $g_{ui} = g_{di} = A d_{ib}^{-\alpha} e^{-\beta X_{ib}}$, where A is the link gain at a reference distance and assumed to be 1, d_{ib} is the distance between MS i and the BS (normalized to the reference distance), α is the path loss constant, X_{ib} is a Gaussian distributed random variable with zero mean and a standard deviation of σ, and $\beta = \ln(10)/10$ is a constant. The outage probability is collected as the number of users in outage divided by the total number of users. It can be seen from the figures that when the number of users (N) is relatively small, the outage probability increases relatively slowly with N; and when N if close to the pole capacity, the outage probability increases significantly with N. From Fig. 2.1 we can see that using criterion 2 results in lower outage probability than using criterion 1. The difference between the outage performance using the two criteria increases as the number of MSs increases.

Note that in a practical cellular system, distributed power control algorithms may be used in both the uplink and the downlink to achieve the target SINRs, such as the distributed and constrained power control introduced in reference [1] (which is introduced in Subsection 1.2.2) and in reference [2]. When performing such algorithms and the exact link gains between the MSs and the BS are unknown, making decisions about which links should be removed during the iterations can be difficult. Several heuristic criteria can be found in [3, 4].

In addition to the maximum transmission power limit, other aspects can also cause outages. In the following two sections, we will look at outages caused by imperfect power control and bursty traffic, and study the required transmission power in each of these cases.

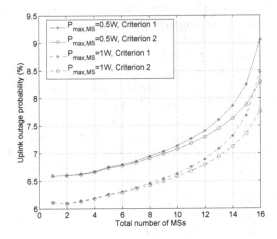

Fig. 2.1 Uplink outage probability for a single cell ($N_p = 16.6$, $\alpha = 4$, $\sigma = 8$dB, $\eta = 10^{-14}$W)

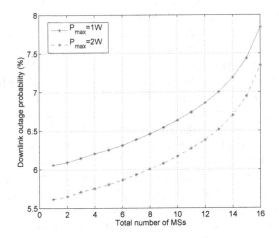

Fig. 2.2 Downlink outage probability for a single cell ($N_p = 16.6$, $\alpha = 4$, $\sigma = 8$dB, $\eta = 10^{-14}$W, $\xi = 1$)

2.3 Effect of Imperfect Power Control

We have introduced an iterative power control scheme in Subsection 1.2.2. Implementing such an iterative power control scheme requires the transmitter and receiver to exchange related information, so that the transmitter knows whether it should increase or decrease the transmission power in the next iteration, and how much the transmission power should be adjusted. In a practical system, the measurement at the receiver may not be accurate, and the transmitted signaling from the receiver to the transmitter can be corrupted by interference and noise. As a result, the transmitter may not adjust its transmission power towards the desired target value, resulting in imperfect power control. In this case, the actual transmission power can be larger or smaller than the desired power. When the transmission power is lower than the desired value, the target SINR of the user cannot be satisfied, causing outage to the communications. In order to protect the user's receiving quality, the target receiving power should be increased, compared to the perfect power control case. The high transmission power increases the co-channel interference in the network, and reduces the system capacity.

When the power control is imperfect, the actual receiving power is the target receiving power multiplied by a random error. Based on [5], the error due to imperfect control is log-normally distributed. Let $\tilde{S}_i$ denote the target receiving power of user i, then the actual receiving power is $\tilde{S}_i e^{\beta Y_i}$, where $\beta = \ln(10)/10$, and Y_i is a normally distributed random variable with zero mean and variance σ_y^2. Larger σ_y indicates larger variations between the target and the actual power. We consider that all the Y_i's are independent and identically distributed. When the actual SINR is below the SINR threshold for user i's transmission, outage occurs to the user. The outage probability is given by

$$P_{out,i} = \text{Pr.} \left\{ \frac{W}{R} \frac{\tilde{S}_i e^{\beta Y_i}}{\sum_{j=1, j\neq i}^{N} \tilde{S}_j e^{\beta Y_j} + \eta} < \gamma^* \right\} \qquad (2.16)$$

for all $i = 1, 2, \ldots, N$. Mathematical analysis and comparison between the required power for perfect and imperfect power control can be found in [6]. Here we use computer simulation results to demonstrate this effect. Fig. 2.3 shows the relationship between the outage probability and the standard deviation of imperfect power control, where the target receiving power is kept the same as that in the prefect power control case. The figure shows that imperfect power control causes communication outages, and the outage probability increases with σ_y. Furthermore, as σ_y increases, the outage probability can increase very significantly towards an unacceptable level, especially when the pole capacity is small.

Fig. 2.4 shows that increasing the target receiving power can effectively reduce the outage probability, but only for a certain range of the outage probability. Beyond this range, increasing the target receiving power has very minor effect on the outage probability. The standard deviation of imperfect power control determines the best outage performance that can be achieved by increasing the target receiving power.

2.4 Adaptive Power and Adaptive Rate

User's traffic often exhibits random active and inactive periods. During an active period, a user generates traffic and transmits to the destination; and during a silent period, the user has no active traffic and does not transmit. Within a network, the number of active users changes randomly, causing random changes in co-channel

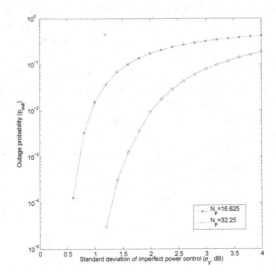

Fig. 2.3 Outage performance vs. standard deviation of imperfect power control ($\eta = 10^{-14}$W and $N = 10$)

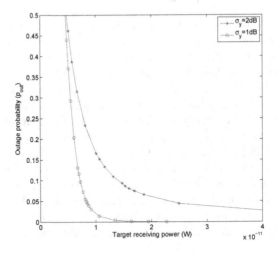

Fig. 2.4 Outage performance vs. target receiving power with imperfect power control ($N_p = 16.625$ and $\eta = 10^{-14}$W)

interference, which affects the resource allocations and system capacity. In this section, we follow some analysis in [7] to look at a simple system with bursty traffic, and study the effect of traffic burstiness on outage, transmission rates, and power allocations.

Consider N_{tot} users, indexed by $i = 1, 2, \ldots, N_{tot}$, all communicating with a common receiver, which, for example, can be the BS in a cellular network. Each user generates bursty traffic. Define a set of binary variables χ_i's. When user i transmits, $\chi_i = 1$; otherwise, $\chi_i = 0$. The transmission rate for an active user i is R_i, and the SINR of its transmission at the BS receiver should satisfy the following condition

$$\frac{W}{R_i} \frac{S}{\sum_{j=1, j \neq i}^{N_{tot}} \chi_j S + \eta} \geq \gamma_i^*. \tag{2.17}$$

Consider homogeneous traffic with $R_i = R$ and $\gamma_i^* = \gamma^*$ for all i. Given R and γ^*, the pole capacity N_p is fixed. Assume there is no maximum transmission power limit. If $N_{tot} \leq N_p$, i.e., the total number of users is less than the pole capacity, there is no outage, since all users can be supported even when they are all active. Let $N_a = \sum_{i=1}^{N_{tot}} \chi_i$ be the random variable representing the number of active users. When $N_a > N_p$, $N_a - N_p$ active users are in outage. As the number of active users changes, the minimum target receiving power changes. Given R and the number of active users, the target receiving power $S = S^*$ for an active user (not in outage) is given by

$$S^*|_{N_a = j} = \frac{\eta R \gamma^*}{W - (\min\{j, N_p\} - 1) R \gamma^*}, \tag{2.18}$$

where $\min\{j, N_p\}$ is due to the fact that when $j > N_p$, only N_p users can be supported. This is referred to as power adaptation, which is to adjust the target receiving power (through transmission power control) so that the required SINR is satisfied for the given transmission rate. The average target receiving power is given by

$$\begin{aligned}
E[S^*] &= \sum_{j=1}^{N_{tot}} S^*|_{N_a = j} \min\{N_p, j\} \Pr.\{N_a = j\} \\
&= \sum_{j=1}^{N_{tot}} \frac{\eta R \gamma^*}{W - (\min\{j, N_p\} - 1) R \gamma^*} \min\{N_p, j\} \Pr.\{N_a = j\}. \tag{2.19}
\end{aligned}$$

Instead of having some users in outage when $N_a > N_p$, the pole capacity can be dynamically changed according to the current number of active users, so that all the active users can be supported. The pole capacity can be changed by adjusting R or γ^*. Assume γ^* is fixed. Reducing R can increase the pole capacity and allow the system to accommodate more users; while larger R is possible when fewer users are active. The channel rate can be changed by varying the spreading factor [8] or using multi-code CDMA technique [9]. Given $N_a = j$, the maximum achieved rate R can be found as

$$R^*|_{N_a = j} = \frac{W}{\gamma^*} \frac{S}{S(j-1) + \eta}. \tag{2.20}$$

As N_a changes, the transmission rate of the active users is adaptively changed. This is referred to as rate adaptation. That is, the target SINR is fixed, and the transmission rate is adjusted so that it is maximized based on the current traffic load and the target receiving power. Given that each user has the same probability of being active, the mean of the transmission rate for a user is given by

$$E[R^*] = \sum_{j=1}^{N_{tot}} R^*|_{N_a=j} \times j \times \Pr.\{N_a = j\}$$

$$= \sum_{j=1}^{N_{tot}} \frac{W}{\gamma^*} \frac{S}{S(j-1)+\eta} \times j \times \Pr.\{N_a = j\}. \qquad (2.21)$$

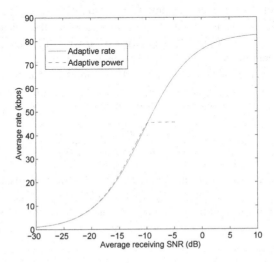

Fig. 2.5 Comparison between adaptive rate and adaptive power ($N_{tot} = 12$, $p_{on} = 0.5$, and $\eta = 10^{-10}$W)

Fig. 2.5 shows the average rate and average receiving SNR (ratio of the target receiving power to the noise power) for both adaptive rate and adaptive power allocations. It is seen that the two systems are approximately the same in low SNR and low rate region, but the adaptive rate system can achieve much higher rate in high SNR region.

- As the number of active users changes randomly, the adaptive rate system allows the users to adjust their transmission rates (and therefore change the pole capacity) based on the current traffic load. When a smaller number of users are active, higher rate can be achieved for each active user; and when a larger number of users are active, a lower rate is served for the users, but none is in outage.
- On the other hand, the adaptive power system fixes the transmission rate, which is independent of the current traffic load. When the number of active users is small, the available network resource is wasted; when the number of active users is large, it has to force some users in outage. In addition, the required power in the power adaptive system can change significantly as the number of active users

changes, making the power adaptive system more likely to have outages than the rate adaptive system when the transmission power is upper bounded.

2.5 Power Allocations in a Multi-cell Network

In this section we study power allocations in a cellular network with multiple cells. Based on the results, we discuss possible approaches to improving the system performance. We use B to represent the total number of BSs, g_{ib} to denote the link gain between BS b and the MS carrying connection i, and $i \in \mathscr{B}_c$ to indicate that connection i is associated to BS c. The main difference between a multi-cell network and a single cell network is that each transmission in a multi-cell network experiences not only interference from the transmissions within the same cell (intra-cell interference), but also that from other cells (inter-cell interference). We consider homogeneous traffic with all the users requiring the same transmission rate R and target SINR γ^*.

Uplink Analysis

In the uplink, power control ensures that all homogeneous connections associated to the same BS have the same power level at the BS receiver input. Let S_b be the target power at the BS receiver input for a connection associated to BS b, and N_b be the total number of connections currently associated to BS b. Then for a given connection associated to BS b, the experienced intra-cell interference for its signal at the BS receiver input is $(N_b - 1)S_b$. The inter-cell interference is from all other cells. For connection i associated to BS c, $c \neq b$, its transmission power is given by S_c/g_{ic}, and the interference level that its transmission causes at the BS b receiver is $S_c g_{ib}/g_{ic}$. Therefore, the total interference that a connection associated to BS b experiences is given by

$$I_b = (N_b - 1)S_b + \sum_{c=1, c \neq b}^{B} S_c \sum_{i \in \mathscr{B}_c} \frac{g_{ib}}{g_{ic}}. \tag{2.22}$$

The SINR at the BS receiver for the connection associated to BS b is given by

$$\gamma_b = \frac{W}{R} \frac{S_b}{I_b + \eta_b}, \tag{2.23}$$

where η_b is the background noise power at the receiver of BS b. When the power control is perfect and all the users transmit at the lowest power, $\gamma_b = \gamma^*$, and

$$\frac{W}{R} \frac{S_b}{I_b + \eta_b} = \gamma^*. \tag{2.24}$$

Replacing I_b in (2.24) with the right-hand side of (2.22) and manipulating, we have the following relationship:

$$S_b - \frac{1}{\frac{W}{R\gamma^*}+1-N_b} \sum_{c=1,c\neq b}^{B} S_c \sum_{i\in\mathcal{B}_c} \frac{g_{ib}}{g_{ic}} = \frac{\eta_b}{\frac{W}{R\gamma^*}+1-N_b}, \tag{2.25}$$

where $b = 1, 2, \ldots, B$. Define

$$\Delta_b = \frac{W}{R\gamma^*} + 1 - N_b = N_p - N_b, \tag{2.26}$$

where $N_p = \frac{W}{R\gamma^*} + 1$ is the pole capacity of a single cell. We can rewrite (2.25) as

$$S_b - \Delta_b^{-1} \sum_{c=1,c\neq b}^{B} S_c \sum_{i\in\mathcal{B}_c} \frac{g_{ib}}{g_{ic}} = \eta_b \Delta_b^{-1}. \tag{2.27}$$

Define vector $\mathbf{S}_u = (S_1, S_2, \ldots, S_B)^T$. The B equations defined by (2.27) (for $b = 1, 2, \ldots, B$) can then be rewritten in a matrix form as

$$(\mathbf{I} - \boldsymbol{\Delta}_u \mathbf{G}_u)\mathbf{S}_u = \boldsymbol{\eta}_u, \tag{2.28}$$

where $\mathbf{I}$ is a $B \times B$ identity matrix, $\boldsymbol{\Delta}_u = \text{diag}(\Delta_1^{-1}, \Delta_2^{-1}, \ldots, \Delta_B^{-1})$, $\mathbf{G}_u$ is a $B \times B$ matrix whose bth row and cth column is given by

$$G_{u,bc} = \begin{cases} 0, & \text{when } b = c \\ \sum_{i\in\mathcal{B}_c} \frac{g_{ib}}{g_{ic}}, & \text{when } b \neq c \end{cases} \tag{2.29}$$

and $\boldsymbol{\eta}_u$ is a column vector whose bth element is given by $\eta_{u,b} = \frac{\eta_b}{\Delta_b}$.

Downlink Analysis

For the downlink transmissions, all the connections associated with the same BS share the BS transmission power. Denote the total transmission power from BS c as P_c, and let P_{ci} be the transmission power from BS c to user i in cell c.

For user i associated with BS c, the signal level at the user's receiver input is $P_{ci}g_{ic}$, the received interference from transmissions for other users associated with the same BS is $\xi(P_c - P_{ci})g_{ic}$, and the interference from a neighboring cell b is $P_b g_{ib}$. Then the SINR of the received signal for the MS is given by

$$\gamma_i = \frac{W}{R} \frac{P_{ci}g_{ic}}{\xi(P_c - P_{ci})g_{ic} + \sum_{b=1,b\neq c}^{B} P_b g_{ib} + \eta_i},$$

$$= \frac{W}{R} \frac{P_{ci}}{\xi(P_c - P_{ci}) + \sum_{b=1,b\neq c}^{B} P_b \frac{g_{ib}}{g_{ic}} + \frac{\eta_i}{g_{ic}}}. \tag{2.30}$$

Letting $\gamma_i = \gamma^*$, the allocated transmission power for each connection can be found as

$$P_{ci} = \frac{1}{W/(\xi \gamma^* R) + 1} \left(P_c + \frac{1}{\xi} \sum_{b=1, b \neq c}^{B} P_b \frac{g_{ib}}{g_{ic}} + \frac{\eta_i}{\xi g_{ic}} \right). \tag{2.31}$$

The total transmission power from BS c for all connections in the cell is given by

$$P_c = \sum_{i \in \mathcal{B}_c} P_{ci}. \tag{2.32}$$

Replacing P_{ci} in (2.32) by the right-hand side of (2.31) and manipulating we have

$$P_c - \frac{\sum_{b=1, b \neq c}^{B} \sum_{i \in \mathcal{B}_c} \frac{g_{ib}}{g_{ic}}}{\Delta_{\xi, c}} P_b = \frac{\sum_{i \in \mathcal{B}_c} \frac{\eta_i}{g_{ic}}}{\Delta_{\xi, c}}, \tag{2.33}$$

where

$$\Delta_{\xi, c} = \xi \left(\frac{W}{\xi \gamma^* R} + 1 - N_c \right) = \xi (N_{\xi, p} - N_c), \tag{2.34}$$

and $N_{\xi, p}$ is the downlink pole capacity of a single cell. The expression in (2.33) gives a set of B linear equations for $c = 1, 2, \ldots, B$.

Define a column vector $\mathbf{P}_d = (P_1, P_2, \ldots, P_B)^T$, a $B \times B$ matrix $\mathbf{G}_d$ whose cth row and bth column element is given by

$$G_{d,cb} = \begin{cases} 0, & \text{if } c = b \\ \sum_{i \in \mathcal{B}_c} \frac{g_{ib}}{g_{ic}}, & \text{if } c \neq b, \end{cases} \tag{2.35}$$

a diagonal matrix $\mathbf{\Delta}_d = \text{diag}(\Delta_{\xi, 1}^{-1}, \Delta_{\xi, 2}^{-1}, \ldots, \Delta_{\xi, B}^{-1})$, and a column vector η_d with the cth element given by

$$\eta_{d,c} = \frac{\sum_{i \in \mathcal{B}_c} \frac{\eta_i}{g_{ic}}}{\Delta_{\xi, c}}. \tag{2.36}$$

Then (2.33) can be rewritten as

$$(\mathbf{I} - \mathbf{\Delta}_d \mathbf{G}_d) \mathbf{P}_d = \eta_d. \tag{2.37}$$

Discussions

The solution to $\mathbf{S}_u$ in (2.28) is the minimum receiving power that satisfies the required SINR of all users in the uplink, and the solution to $\mathbf{P}_d$ in (2.37) is the minimum BS transmission power in order to support all users in the downlink. These two equations can be combined in a common form as

$$(\mathbf{I} - \mathbf{\Delta} \mathbf{G}) \mathbf{P} = \eta, \tag{2.38}$$

where $\mathbf{G} = \mathbf{G}_u$, $\mathbf{P} = \mathbf{S}_u$, $\boldsymbol{\Delta} = \boldsymbol{\Delta}_u$, and $\eta = \eta_u$ for the uplink, and $\mathbf{G} = \mathbf{G}_d$, $\mathbf{P} = \mathbf{P}_d$, $\boldsymbol{\Delta} = \boldsymbol{\Delta}_d$, and $\eta = \eta_d$ for the downlink. When there is a feasible solution to $\mathbf{P}$, i.e., all elements in $\mathbf{P}$ are non-negative (assume there is no maximum transmission power limit), the required SINR is achievable, and all the users can be simultaneously supported. From the analysis in previous sections we know that the power distribution is feasible if and only if the dominant eigenvalue of $\boldsymbol{\Delta}\mathbf{G}$, denoted as $\rho(\boldsymbol{\Delta}\mathbf{G})$, is less than 1. When any element of $\mathbf{P}$ is less than 0, the target SINR or required transmission rate should be reduced, if the same number of users are to be supported. Otherwise, some users should be in outage so that the remaining users can be served with the required rate and SINR.

When each BS is supporting at least one connection, we can see that $\boldsymbol{\Delta}\mathbf{G}$ is non-negative (element-wise) for both the uplink and the downlink. Furthermore, since i) $\boldsymbol{\Delta}$ is a diagonal matrix with all the diagonal elements larger than zero, and ii) the diagonal elements of $\mathbf{G}$ are all zero and all other elements in $\mathbf{G}$ are greater than zero, we can conclude that the matrix given by $\boldsymbol{\Delta}\mathbf{G}$ is also irreducible. For such matrices, the Perron-Frobenius Theorem [10] indicates that increasing any entry of $\boldsymbol{\Delta}\mathbf{G}$ may increase $\rho(\boldsymbol{\Delta}\mathbf{G})$. Therefore, larger elements of $\boldsymbol{\Delta}\mathbf{G}$ lead to a higher possibility that $\rho(\boldsymbol{\Delta}\mathbf{G}) > 1$, and a higher chance that $\mathbf{P}$ is infeasible. When the transmission rate and SINR requirements of the connections are given, large values in $\boldsymbol{\Delta}\mathbf{G}$ may be due to large values in these two matrices. A large element in $\boldsymbol{\Delta}$ may be due to a large number of users in a cell (large N_c) or small single cell pole capacity (large $\gamma^* R$), and a large element in $\mathbf{G}$ means high normalized link gain (g_{ib}/g_{ic} for $i \in \mathscr{B}_c$), which is resulted from poor transmission conditions — weak desired link and/or strong interfering link. If it is possible to control the elements in $\boldsymbol{\Delta}$ and $\mathbf{G}$ through different techniques so that $\rho(\boldsymbol{\Delta}\mathbf{G})$ can be reduced, then the power allocation problem may be changed from infeasible to feasible. In the next section we introduce the method of using soft handoff to achieve this objective, and in Section 4.1 the technique of using multihop relaying is introduced to improve the relative link gains so that to improve the power allocation feasibility as well as the network performance.

2.6 Soft Handoff and Power Allocations

When an MS is moving across the boundary of two cells, it should handoff from one BS to another BS. In a wireless cellular network where neighboring radio cells use different frequency channels, the MS has to disconnect from the previous BS before connecting to the new BS. This is called hard handoff (HHO). During the time period when the handoff is performed, communication outage occurs and packets may be lost. The CDMA-based cellular networks allow soft handoff (SHO). Since all the cells share the same spectrum band, there is no need to break the connection from the previous BS before establishing a connection to the new BS. During the SHO period, an MS can simultaneously connect to multiple BSs. Performing SHO can make data transmissions smoother than using HHO. The multiple BSs that an MS can be connected to during SHO form a set, referred to as the active BS set of the

MS. In the uplink, the transmitted signal of an MS can be received simultaneously by all BSs in the active set; and in the downlink, an MS can simultaneously receive the same signals from all the BSs in the active set. Compared with using HHO, using SHO allows the MS to take advantage of the diversity provided by multiple BSs in the active set and always connect to the "best" BS [11]. By taking advantage of this property, transmission power of the MSs and BSs may be reduced. As the CDMA network is interference limited, minimizing the transmission power is directly related to improving the network capacity.

In the uplink, if multiple BSs send power control commands to an MS, the decision for increasing the transmission power at the MS is made only if all the BSs in the active set require it to increase the power. That is, the transmission power from the MS is only required to guarantee sufficient SINR at one BS in the active set. This ensures that the transmission power of the MS is always the minimum. Let P_m represent the transmission power of MS m, then

$$P_m = \min_{b \in \mathscr{S}_b} S_b/g_{mb}, \tag{2.39}$$

where $\mathscr{S}_b$ is the active BS set of MS m, and S_b is the target receiving power at BS b. Assuming S_b's are known for all b, P_m can be found. With the objective to minimize the MS transmission power, the "best" BS for MS m is given by

$$b^* = \arg \min_{b \in \mathscr{S}_b} S_b/g_{mb}. \tag{2.40}$$

As an example, consider that an MS is located in the middle of BSs 1 and 2, and is connected to BS 1 if HHO is performed. If SHO is performed, then both BSs 1 and 2 are in its active set. The transmission power from the MS using HHO is S_1/g_{m1}, and using SHO is $\min\{S_1/g_{m1}, S_2/g_{m2}\}$. When $S_1 = S_2$, the ratio of the transmission power using HHO to that using SHO is

$$\frac{S_1/g_{m1}}{\min\{S_1/g_{m1}, S_2/g_{m2}\}} = \frac{1/g_{m1}}{\min\{1/g_{m1}, 1/g_{m2}\}}. \tag{2.41}$$

Using the same channel model with path loss and log-normally distributed shadowing as in Sections 2.2, and assuming $d_{m1} = d_{m2}$ and independent shadowing effect between the MS to the two BSs, Fig. 2.6 shows the average of the power ratio, where the x-axis is the standard deviation of the shadowing. From the figure we can see that the transmission power can be largely reduced by using SHO. Furthermore, the effect of this becomes more significant when there are larger variations in channel fading. That means, using SHO as defined above can get more benefit in highly fading channels.

From (2.39) we can see that whether a BS in the active set can become the "best" BS of MS m depends on both the target receiving power of the connection at the BS and the link gain between MS m and the BS. From the analysis in previous sections we know that given the transmission rate and SINR requirements, the value of S_b depends on the number of MSs associated to BS b, or N_b. The values of N_b count

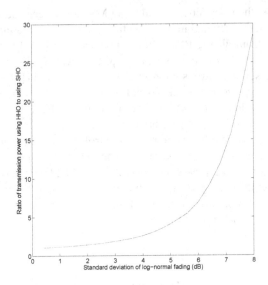

Fig. 2.6 Ratio of transmission power using HHO to transmission power using SHO in uplink: single MS case

both the HHO MSs associated to BS b and the SHO MSs having BS b as their "best" BS. In a practical system, performing optimum SHO and power control in order to minimize the transmission power of all MSs can be difficult, since this requires the knowledge of the "best" BSs of all SHO MSs, and the power allocations of the MSs in all the cells should be jointly performed.

Next we consider a simple SHO decision. Each active MS can communicate with the two nearest BSs, and always chooses to connect to the one to which it has better link gain. If BSs b and c are the two nearest BSs for MS m, then $i \in \mathscr{B}_c$ when $g_{ic} \geq g_{ib}$, and $i \in \mathscr{B}_b$ otherwise. Using this criterion, the association between the MSs and the BSs does not depend on the traffic load in each cell, and the analysis in Section 2.5 can be used to find the required transmission power for each MS. We consider that all the users transmit at the same rate, and find the maximum transmission rate that can be supported to the users. For comparison, we also consider a systems with HHO only, where all MSs communicate directly with their nearest BSs. Note that when the channels experience random fading, the MSs performing SHO can switch between different BSs more dynamically. A simple algorithm as shown in Algorithm 1 is used to find the maximum transmission rate for given link gains. We start from a small rate, and find the transmission power using the analytical model developed in Section 2.5. A variable UP is initially reset to zero. If the power allocation is feasible, the rate is doubled. This is repeated until the power allocation is infeasible, when UP is set to 1. Record this rate as $2R_1$. The maximum rate is then between R_1 and $2R_1$. The transmission rate is then returned to R_1, and a step size is initialized to $R_1/2$. Starting from this point, the rate is either increased or decreased by a step size after each iteration, depending on whether the power allocation is feasible in the current iteration, and then the step size is halved. This process is repeated until the step size is very small and the change that it causes to the rate can be neglected. In the algorithm, R_{min} is the minimum step size, which is set to 1bps in generating the numerical results. We simulate a two-dimensional cellular network, which consists of 19 hexagonal cells as shown in Fig. 2.7, where cell 1 is the center cell, cells 2 to 7 are the six first tier cells, and cells 8 to 19 are the second tier cells. A large set of the rates are obtained based on randomly generated link gains and then averaged. We consider distance related path loss and independent log-normal fading. The link gain model between the MSs and the BSs is the same as in Section 2.2. The CDMA bandwidth $W = 5$MHz, the path loss exponent is $\alpha = 4$, the standard deviation for shadowing is 8dB, the channel orthogonality factor for the downlink is $\xi = 0.5$, and the target SINR for the users' traffic is 6.8dB. Figs. 2.8 and 2.9 show the transmission rate that can be supported when the number of MSs changes in the uplink and the downlink, respectively. From these figures we can observe that using SHO can provide higher transmission rate than using HHO, and this is consistent for both the uplink and the downlink.

Algorithm 1 Finding the maximum transmission rate

1: $F = 0$, UP=0, and $R = 1$bps.
2: **while** $F = 0$ **do**
3:      Find target receiving power and transmission power
4:      **if** UP=0 **then**
5:          **if** Solution is feasible **then**
6:              $R = 2R$;
7:          **else**
8:              $R = R/2$, UP=1, and $\delta_R = R/2$.
9:          **end if**
10:     **else**
11:         **if** Solution is feasible **then**
12:             $R = R + \delta_R$;
13:         **else**
14:             $R = R - \delta_R$;
15:         **end if**
16:     **end if**
17:     **if** $\delta_R < R_{min}$ **then**
18:         $F = 1$;
19:         Record the value of R;
20:     **else**
21:         $\delta_R = \delta_R/2$;
22:     **end if**
23: **end while**

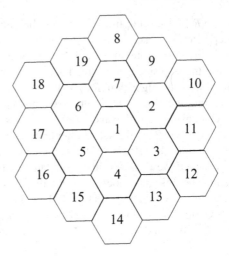

Fig. 2.7 Cell layout

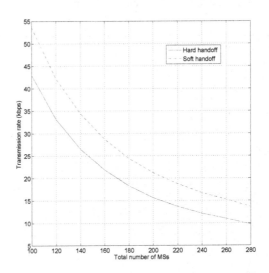

Fig. 2.8 Transmission rates for the uplink using SHO and HHO

References

1. Grandhi SA, Zander J, Yates RD (1995) Constrained power control. Wireless Personal Communications 1(4): 257-270.
2. Jiang H, Wang P, Zhuang W, Shen X (2007) An interference aware distributed resource management scheme for CDMA-based wireless mesh backbone. IEEE Transactions on Wireless Communications 6(12): 4558-4567.

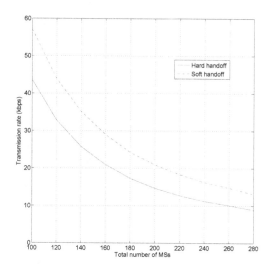

Fig. 2.9 Transmission rates for the downlink using SHO and HHO

3. Andersin M, Rosberg Z, Zander J (1996) Gradual removals in cellular PCS with constrained power control and noise. ACM/Balzer Wireless Networks Journal 2(1): 27-43.
4. Berggren F, Jantti R, Kim SL (2001) A generalized algorithm for constrained power control with capability of temporary removal. IEEE Transactions on Vehicular Technology 50(6): 1604-1612.
5. Viterbi AJ, Viterbi AM, Zehavi E (1993) Performance of power controlled wideband terrestrial digital communication. IEEE Transactions on Communications (41): 559-569.
6. Zhao D, Shen X, Mark JW (2001) Uplink power distribution in MC-CDMA systems supporting heterogeneous services. Proc. IEEE Global Telecommunications Conference (GLOBE-COM) 5: 3040-3044.
7. Xu L, Shen X, Mark JW (2001) Performance analysis of adaptive rate and power control for data service in DS-CDMA systems. Proc. IEEE Globecom: 627-631.
8. I CL, Sabnani KK (1995) Variable spreading gain CDMA with adaptive control for true packet switching wireless DS/CDMA networks. Proc. IEEE International Conference on Communications (ICC): 725-730.
9. I CL, Gitlin RD (1995) Multi-code CDMA wireless personal communications networks. Proc. IEEE International Conference on Communications (ICC): 1060-1064.
10. Varga RS (1962) Matrix iterative analysis, Chapter 2. Prentice-Hall, Englewood Cliffs, N.J.
11. Zhao D, Shen X, Mark JW (2006) QoS guarantee and optimum power distribution for soft handoff connections in cellular CDMA downlink. IEEE Transactions on Wireless Communications 5(4): 910-919.

Chapter 3
Wireless Networks with Two-hop Relaying

Abstract In this chapter, we look at the performance of transmission links with two-hop relaying. Each link has two end nodes, one source node and one destination node, and one or multiple relay nodes. The direct link between the end nodes may or may not exist. The relay nodes can use either amplify-and-forward (AF) or decode-and-forward (DF) when forwarding received signals. In one-way relaying, the relay node receives from one end node and then forwards to the other end node; and in two-way relaying, the relay node combines the received signals from both the end nodes and forwards simultaneously to the two end nodes. Relationship between link capacity and node transmission power for different combinations of the link and transmission conditions is studied.

keywords: Two-hop relay, capacity, signal-to-interference-plus-noise ratio, amplify-and-forward, decode-and-forward, network coding, two-way relay.

3.1 Multihop Relaying

Using relaying can break a direct transmission between a source node and a destination node into multiple shorter links and thus reduce the total required transmission power. For a simple example, let d_{sd} represent the distance between the source and the destination nodes, and S be the minimum receiving power required in order for the destination to recover the desired signal under given transmission conditions. Without considering channel fading, the minimum transmission power from the source node is given by $P_{sd} = d_{sd}^{\alpha}S$, where α is the path loss exponent. With a relay station between the two nodes, the direct transmission is broken into two hops. In the first hop, the source node transmits to the relay node; and in the second hop, the relay node forwards the signal that it receives from the source to the destination node. In order for the signal to be correctly transmitted along each hop, the minimum transmission power of the source node is given by $P_{sr} = d_{sr}^{\alpha}S$, and the minimum transmission power of the relay node is given by $P_{rd} = d_{rd}^{\alpha}S$, where d_{sr} and d_{rd}, respectively, are the distance between the source and the relay nodes and the distance between the relay and the destination nodes. When the relay node is located exactly in the middle point along the line between the source and destination nodes, $d_{sr} = d_{rd} = d_{sd}/2$, and the total transmission power of the source and the relay nodes is $2(d_{sd}/2)^\alpha S$. For a typical outdoor environment with $\alpha = 4$, the total transmission power using relay is only 1/8 of the transmission power without relay.

Using relay can extend the communication distance between the source and the destination. Given the transmission power of the source node P_{sd}, the maximum communication distance between the source and destination nodes is $d_{sd} = (P_{sd}/S)^{1/\alpha}$. When a relay node is used, and suppose the transmission power of the source plus that of the relay node is the same as the transmission power of the source without using relay, i.e., $P_{sr} + P_{rd} = P_{sd}$, then the communication distance between the source and destination nodes is maximized when the relay is located in the middle of the two end nodes and equal to $2 \times (1/2)^{1/\alpha} \approx 1.68$ times of the original communication distance without relaying.

In addition to saving transmission power and extending communication distances, using relay can also bring other benefits to wireless communications by exploiting the diversity gains provided by the direct link and the relay links. On the other hand, using relay brings other issues, such as time synchronization and relay station selection, and it makes resource management and performance analysis more complicated [1, 2]. If a relay node is equipped with only a single radio, it should switch between receiving from the upstream node and transmitting to the downstream node. This not only requires time synchronization among the nodes, but may also reduce the overall throughput from the source to the destination. In two hop transmissions, the source-to-destination throughput is limited by the weaker hop. Selecting the relay station can be an important issue, because it affects the transmission conditions of both hops. The problems become more complicated for multihop relaying.

This section focuses on two-hop links only. Depending on the techniques used at the relay nodes to process the received data, there can be decode-and-forward (DF) and amplify-and-forward (AF). When using DF, upon receiving the signal from the source node, the relay node should decode the signal first, re-encode it, and then forward to the destination. When using AF, the relay node simply amplifies the received signal together with interference and noise, and then forwards it to the destination. Communications between the two end nodes can be either unidirectional or bidirectional. In the former case, the relay node always receives signals from one end node, and then forwards to the other end node. In the latter case, the relay node should forward data in both directions. It may forward data in one direction at a time, in which case the bidirectional communication is equivalent to two unidirectional communications. Alternatively, the relay node in a bidirectional communication can combine the received data from the two end nodes, and then transmit the mixed data simultaneously to both the end nodes. This is referred to as two-way relay, which is a simple version of network coding applied in wireless networks. In contrast, the conventional relaying that restricts the relay node to forward data to one end node at a time is referred to as one-way relay. Unidirectional communications only need one-way relaying, and bidirectional communications can use either one-way or two-

(a) Unidirectional communication, one–way relay

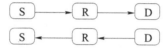

(b) Bidirectional communication, one–way relay

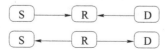

(c) Bidirectional communication, two–way relay

Fig. 3.1 One-way relaying and two-way relaying

way relaying. This is illustrated in Fig. 3.1, where R is the relay node, and S and D are the end nodes.

In the remaining part of this chapter, we look at the capacity performance for the two-hop links using AF and DF for unidirectional and bidirectional communications. We assume that the link gains between two nodes are the same in both directions.

3.2 Performance for Amplify-and-Forward

3.2.1 One-Way Communications and One Relay Node

Without direct transmissions

We consider a two-hop transmission link. Each source-to-destination transmission takes two time slots. In the first time slot, the source transmits to the relay node; and in the second time slot, the relay node transmits to the destination. Fig. 3.2 shows the timeline activities of the nodes, where "S", "R" and "D", respectively, represent the source, relay, and destination nodes, and "TX" and "RX", respectively, represent "transmit" and "receive". (The same notations are used for other diagrams in this chapter.) Define g_{sr}, g_{rd}, and g_{sd}, respectively, as the link gains between the source and the relay, the relay and the destination, and the source and the destination. In the first time slot, if the transmission power of the source is P_s, the receiving power at the relay is $P_s g_{sr}$. In the second time slot, the relay amplifies the received signal as well as the interference and noise by λ times. Let η_r be the total power of the noise

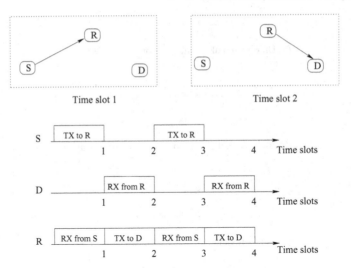

Fig. 3.2 Two-hop relay without direct transmissions

and interference at the relay node. The transmission power of the desired signal from the relay node is $P_s g_{sr} \lambda$, and that of the interference and noise is $\eta_r \lambda$. At the destination receiver, the power of the desired signal is given by $P_s g_{sr} \lambda g_{rd}$, and that of the interference and noise is $\eta_r \lambda g_{rd} + \eta_d$, where the first term is the power of the noise and interference from the relay node, and the second term is the power of noise and interference at the destination node. Overall, the SINR at the destination receiver is given by

$$\gamma_1 = \frac{P_s g_{sr} \lambda g_{rd}}{\eta_r \lambda g_{rd} + \eta_d}. \tag{3.1}$$

Let P_r represent the total transmission power of the relay. We have

$$P_r = (P_s g_{sr} + \eta_r)\lambda, \tag{3.2}$$

from which λ can be found as

$$\lambda = \frac{P_r}{P_s g_{sr} + \eta_r}. \tag{3.3}$$

Substituting λ in (3.3) into (3.1) and rearranging we have

$$\gamma_1 = \frac{P_s P_r g_{sr} g_{rd}}{P_r \eta_r g_{rd} + P_s \eta_d g_{sr} + \eta_d \eta_r}. \tag{3.4}$$

Define $\gamma_r = P_s g_{sr}/\eta_r$, which is the SINR at the relay node. Further define $\gamma_d = P_r g_{rd}/\eta_d$, which is the SINR at the destination, assuming all the relay transmission power is for the desired signal. After dividing both the numerator and the denominator on the right-hand side of (3.4) by $\eta_r \eta_d$, we have

$$\gamma_1 = \frac{\frac{P_s g_{sr}}{\eta_r} \frac{P_r g_{rd}}{\eta_d}}{\frac{P_r g_{rd}}{\eta_d} + \frac{P_s g_{sr}}{\eta_r} + 1} = \frac{\gamma_r \gamma_d}{\gamma_d + \gamma_r + 1}. \tag{3.5}$$

For Gaussian channel, the source-to-destination (S-D) link capacity is given by

$$R_{AF,1} = \frac{1}{2} \log_2(1 + \gamma_1), \tag{3.6}$$

where the fraction $\frac{1}{2}$ is due to the fact that the source can only transmit in every other time slot.

Dividing both the numerator and the denominator of the expression for γ_1 on the right-hand side of (3.5) by γ_r and combining with (3.6), we can rewrite the S-D link capacity as

$$R_{AF,1} = \frac{1}{2} \log_2 \left(1 + \frac{\gamma_d}{1 + \frac{\gamma_d + 1}{\gamma_r}} \right). \tag{3.7}$$

Similarly, dividing both the numerator and the denominator of the expression for γ_1 on the right-hand side of (3.5) by γ_d and combining with (3.6), we can rewrite the S-D link capacity as

$$R_{AF,1} = \frac{1}{2} \log_2 \left(1 + \frac{\gamma_r}{1 + \frac{\gamma_r + 1}{\gamma_d}} \right). \tag{3.8}$$

Letting $\gamma_r \to \infty$ in (3.7) and $\gamma_d \to \infty$ in (3.8) we can find the upper bounds of the S-D link capacity as

$$R_{AF,1} \le \frac{1}{2} \log_2 (1 + \gamma_d) = \frac{1}{2} \log_2 \left(1 + \frac{P_r g_{rd}}{\eta_d} \right), \tag{3.9}$$

$$R_{AF,1} \le \frac{1}{2} \log_2 (1 + \gamma_r) = \frac{1}{2} \log_2 \left(1 + \frac{P_s g_{sr}}{\eta_r} \right), \tag{3.10}$$

where (3.9) is the capacity achieved by the relay-to-destination hop, and (3.10) is the capacity achieved by the source-to-relay hop. The S-D capacity is limited by the weaker hop. Given P_s, the effect of increasing P_r on the S-D link capacity is asymptotically upper bounded by the source-to-relay capacity; and similarly, given P_r, the effect of increasing P_s on the S-D link capacity is asymptotically upper bounded by the relay-to-destination capacity. In order to increase the S-D capacity, both P_s and P_r should be increased proportionally.

When multiple links share the same frequency channel, η_r and η_d include both the background noise and interference from other simultaneous transmissions. In such a case, transmissions from different source and relay nodes cause interference to each other. Based on different objectives, such as minimizing the maximum transmission power of all the nodes in the network, maximizing the total throughput, or providing fair throughout among different links, power allocations can be performed accordingly. Some of these problems are formulated and solved in [3].

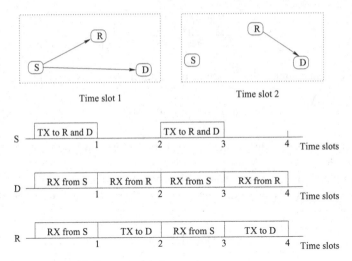

Fig. 3.3 Two-hop relay with direct transmissions

With direct transmissions

When the direct link between the source and the destination exists, the source transmits in the first time slot, and both the relay and the destination can receive from it. In the second time slot, the relay node transmits and the destination receives. The destination, after receiving from the source in the first time slot and from the relay node in the second time slot, combines the two copies of the received signal and decodes it.

For the direct transmission from the source to the destination, the SINR of the signal at the destination is $\frac{P_s g_{sd}}{\eta_d}$. The expression for the SINR of the forwarding link at the destination is the same as in (3.4). If maximum ratio combining (MRC) is used at the destination, the overall SINR for decoding the desired signal at the destination is a sum of the two SINRs, and is given by

$$\gamma_2 = \frac{P_s g_{sd}}{\eta_d} + \frac{P_s P_r g_{sr} g_{rd}}{P_r \eta_r g_{rd} + P_s \eta_d g_{sr} + \eta_d \eta_r}, \qquad (3.11)$$

and the S-D link capacity is given by

$$R_{AF,2} = \frac{1}{2} \log_2(1 + \gamma_2). \qquad (3.12)$$

Obviously, $R_{AF,2} > R_{AF,1}$, because the direct link transmission contributes to the SINR at the destination. Increasing P_s monotonically increases $R_{AF,2}$, because the capacity of the direct link increases with P_s. Dividing the second term on the right-hand side of (3.11) by P_r in both the numerator and the denominator, we have

$$R_{AF,2} = \frac{1}{2} \log_2 \left(1 + \frac{P_s g_{sd}}{\eta_d} + \frac{P_s g_{sr} g_{rd}}{\eta_r g_{rd} + \frac{P_s}{P_r} \eta_d g_{sr} + \frac{\eta_d \eta_r}{P_r}} \right), \tag{3.13}$$

from which we can see that the effect of increasing P_r on $R_{AF,2}$ is asymptotically upper bounded. For an extreme case, when $P_r \to \infty$, we can find an upper bound of $R_{AF,2}$ as

$$R_{AF,2} \le \frac{1}{2} \log_2 \left(1 + \frac{P_s g_{sd}}{\eta_d} + \frac{P_s g_{sr}}{\eta_r} \right). \tag{3.14}$$

When the transmission quality of the direct link is significantly better than that of the forwarding link, the first term is much larger than the second term on the right-hand side of (3.11), and the relay power has little effect on the achievable S-D link capacity.

3.2.2 One-Way Relay and Multiple Relay Nodes

The analysis in the previous section can be extended to a more general scenario when multiple relay nodes are available to forward traffic from the source to the destination. Let M be the number of relay nodes. In order to realize orthogonal transmissions, only one node (source or relay) can transmit in each time slot. Therefore, it takes $M + 1$ time slots to complete one packet transmission from the source to the destination, one time slot for the source to transmit, and one for each of the M relay nodes to transmit. Upon receiving all the transmissions, the destination combines the signals using MRC. Similar to the derivations in the previous subsections, we can find the SINR at the destination as

$$\gamma_M = \frac{P_s g_{sd}}{\eta_d} + \sum_{m=1}^{M} \frac{P_s P_{r,m} g_{sr,m} g_{rd,m}}{P_{r,m} \eta_{r,m} g_{rd,m} + P_s \eta_d g_{sr,m} + \eta_d \eta_{r,m}}, \tag{3.15}$$

where $P_{r,m}$ is the transmission power of the mth relay node, $g_{sr,m}$ and $g_{rd,m}$, respectively, represent the link gain between the source and the mth relay node and the link gain between the mth relay node and the destination, and $\eta_{r,m}$ is the interference and noise power at the mth relay node. The S-D link capacity is given by

$$R_{AF,M} = \frac{1}{M+1} \log_2 (1 + \gamma_M). \tag{3.16}$$

In both (3.15) and (3.16) we assume that the direct link between the source and the destination exists.

Although having all the M relay nodes participate in the transmissions achieves full diversity order, the factor $1/(M+1)$ has a large adverse effect on the overall throughput. One method to solve this problem is to select one relay node among the M available relay nodes, and use the selected one only for forwarding traffic. This is referred to as Selection Amplify-and-Forward (S-AF) [4], which requires two time

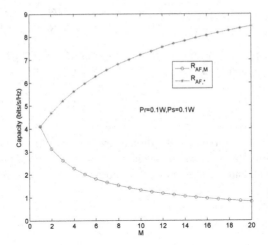

Fig. 3.4 Capacity vs. power for one-way relaying using AF and M relay nodes ($\alpha = 3$, and $\sigma = 4$dB)

slots to complete one packet transmission from the source to the destination. When the m^*th relay node is used, the S-D capacity is given by

$$R_{AF,*} = \frac{1}{2} \log_2 \left(1 + \frac{P_s g_{sd}}{\eta_d} + \frac{P_s P_{r,m^*} g_{sr,m^*} g_{rd,m^*}}{P_{r,m^*} \eta_{r,m^*} g_{rd,m^*} + P_s \eta_d g_{sr,m^*} + \eta_d \eta_{r,m^*}} \right). \quad (3.17)$$

The best relay node that maximizes $R_{AF,*}$ is given by

$$m^* = \arg\max_{m} \frac{P_s P_{r,m} g_{sr,m} g_{rd,m}}{P_{r,m} \eta_{r,m} g_{rd,m} + P_s \eta_d g_{sr,m} + \eta_d \eta_{r,m}}. \quad (3.18)$$

We use computer simulation to generate $R_{AF,M}$ and $R_{AF,*}$ based on the same channel model as in Section 2.2, where the link gains include path loss and log-normally distributed channel fading. All the link gains are independent of each other. Fig. 3.4 shows that $R_{AF,M}$ decreases with M, while $R_{AF,*}$ increases with M. When M is large, using S-AF can significantly improve the capacity. The price for the higher capacity in S-AF is more complicated implementation in order to select the best relay node. Reference [4] provides more details of the implementation.

3.2.3 Two-Way Relay and One Relay Node

When both the source and destination nodes (referred to as the end nodes) have data to transmit to one another, and a relay node is in between for forwarding their transmissions, it takes a minimum of two time slots for the two end nodes to exchange a pair of packets (one in each direction). In the first time slot, the source node transmits packet X_1 and the destination node transmits packet X_2 simultaneously to the relay node. Let P_s and P_d, respectively, be the transmission power of the source and destination. The receiving power at the relay node is $P_s g_{sr}$ for X_1 and $P_d g_{dr}$ for X_2. Upon receiving the signals, the relay node amplifies the mixed signals together with interference and background noise by λ times, and forwards to both the end nodes in the second time slot. The power for each portion in the transmitted mixed signal is given by

$$\text{for } X_1: \qquad P_s g_{sr} \lambda, \qquad (3.19)$$

$$\text{for } X_2: \qquad P_d g_{dr} \lambda, \qquad (3.20)$$

$$\text{for interference and noise: } \eta_r \lambda. \qquad (3.21)$$

The total transmission power of the relay node is given by

$$P_r = \lambda (P_s g_{sr} + P_d g_{dr} + \eta_r), \qquad (3.22)$$

from which we can solve λ as

$$\lambda = \frac{P_r}{P_s g_{sr} + P_d g_{dr} + \eta_r}. \qquad (3.23)$$

The relay transmitted signal reaches both the source and the destination nodes. The power for each portion of the mixed signal at the source node receiver is as follows:

$$\text{for } X_1: \qquad P_s g_{sr} \lambda g_{rs}, \qquad (3.24)$$

$$\text{for } X_2: \qquad P_d g_{dr} \lambda g_{rs}, \qquad (3.25)$$

$$\text{for interference and noise: } \eta_r \lambda g_{rs}. \qquad (3.26)$$

Similarly, the power for each portion of the mixed signal at the destination node receiver is as follows:

$$\text{for } X_1: \qquad P_s g_{sr} \lambda g_{rd}, \qquad (3.27)$$

$$\text{for } X_2: \qquad P_d g_{dr} \lambda g_{rd}, \qquad (3.28)$$

$$\text{for interference and noise: } \eta_r \lambda g_{rd}. \qquad (3.29)$$

The decoding at the two end nodes at the second time slot can exploit the fact that each of them already knows its prior transmitted signal in the first time slot. Assuming each of the end nodes knows λ and the link gain between itself and the relay node, the source node can successfully remove the portion of X_1 from its received signal, and the destination node can successfully remove the portion of X_2 from its received signal. After removing the portion of X_1, the SINR for the source node to recover X_2 is given by

$$\gamma_{DS} = \frac{P_d g_{dr} g_{rs} \lambda}{\eta_r g_{rs} \lambda + \eta_s}, \qquad (3.30)$$

where η_s is the noise power at the source node receiver, and the destination-to-source (D-S) link capacity is given by

$$R_{DS2w,AF} = \frac{1}{2} \log_2 \left(1 + \frac{P_d g_{dr} g_{rs} \lambda}{\eta_r g_{rs} \lambda + \eta_s} \right). \tag{3.31}$$

Similarly, after removing the portion of X_2, the SINR for the destination node to recover X_1 is given by

$$\gamma_{SD} = \frac{P_s g_{sr} g_{rd} \lambda}{\eta_r g_{rd} \lambda + \eta_d}, \tag{3.32}$$

based on which, the S-D link capacity is given by

$$R_{SD2w,AF} = \frac{1}{2} \log_2 \left(1 + \frac{P_s g_{sr} g_{rd} \lambda}{\eta_r g_{rd} \lambda + \eta_d} \right). \tag{3.33}$$

The capacity formulas in (3.31) and (3.33) are derived in [5].

By submitting λ in (3.23) into the expressions of γ_{DS} in (3.30) and γ_{SD} in (3.32) and rearranging we have

$$\gamma_{DS} = \frac{P_d P_r g_{dr} g_{rs}}{P_r \eta_r g_{rs} + P_s \eta_s g_{sr} + P_d \eta_s g_{rd} + \eta_s \eta_r}, \tag{3.34}$$

and

$$\gamma_{SD} = \frac{P_s P_r g_{sr} g_{rd}}{P_r \eta_r g_{rd} + P_s g_{sr} \eta_d + P_d g_{dr} \eta_d + \eta_r \eta_d}. \tag{3.35}$$

From (3.34) and (3.35) one can see that:

- Increasing P_r can increase the SINRs in both directions, but the effect of this is asymptotically upper bounded. When $P_r \to \infty$, and $P_s, P_d \ll P_r$, we have

$$\lim_{P_r \to \infty} \gamma_{DS} = \frac{P_d g_{dr}}{\eta_r}, \tag{3.36}$$

and

$$\lim_{P_r \to \infty} \gamma_{SD} = \frac{P_s g_{sr}}{\eta_r}. \tag{3.37}$$

That is, the capacity in each direction is upper bounded by the capacity of the respective first hop transmission.

- Increasing the transmission power of the end nodes has a contradictory effect on the S-D and D-S link capacities. Increasing P_s increases γ_{SD} but decreases γ_{DS}. Similarly, increasing P_d increases γ_{DS} but decreases γ_{SD}. For an extreme case, when $P_d \to \infty$, and $P_s, P_r \ll P_d$, we have

$$\lim_{P_d \to \infty} \gamma_{DS} = \frac{P_r g_{rs}}{\eta_s}, \tag{3.38}$$

and

$$\lim_{P_d \to \infty} \gamma_{SD} = 0. \tag{3.39}$$

Likewise, when $P_s \to \infty$, and $P_d, P_r \ll P_s$, we have

$$\lim_{P_s \to \infty} \gamma_{DS} = 0 \tag{3.40}$$

and

$$\lim_{P_s \to \infty} \gamma_{SD} = \frac{P_r g_{rd}}{\eta_d}. \tag{3.41}$$

Both (3.38) and (3.41) indicate that the capacity in each direction is limited by the capacity of the respective second hop transmissions.

The two-way communication link with relay described in this section is a simple application of network coding in wireless networks. The relay node does not forward each received packet separately, but broadcasts to both the end nodes a function of the packets that it has received from both the end nodes. The two end nodes each recover their desired packet based on their knowledge about the other packet included in the received signal. Since the relay node does not try to decode the received signals, but only amplifies and forwards the received analog signals, this type of network coding is also referred to as physical layer network coding. Another application of network coding in multihop relaying is introduced in Subsection 3.3.2.

3.3 Performance for Decode-and-Forward

3.3.1 One-Way Communications

Without Direct Link

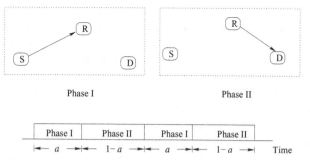

Fig. 3.5 One-way relay using DF

As shown in Fig. 3.5, consider a time window with a fixed duration that is normalized to 1. The window is divided into two periods, each corresponding to one transmission phase in the two-hop transmissions from the source to the destination. The first phase lasts for a, during which the source transmits to the relay; and the second phase lasts for $1 - a$, during which the relay transmits the signal that it decoded in the first phase to the destination. Without a direct link between the source and the destination, the S-D link capacity, $R_{DF,1}$, is given by

$$R_{DF,1} = \min\{R_1, R_2\}, \tag{3.42}$$

$$R_1 = a\log_2\left(1 + \frac{g_{sr}P_s}{\eta_r}\right), \tag{3.43}$$

$$R_2 = (1 - a)\log_2\left(1 + \frac{g_{rd}P_r}{\eta_d}\right), \tag{3.44}$$

where R_1 represents the maximum rate at which the relay can reliably decode the source message, and R_2 represents the maximum rate at which the destination can reliably decode the message from the relay. The two-hop link capacity is limited by the capacity of the weaker hop.

Given the transmission power of the source and relay nodes and the link gains, the optimum value of a that maximizes the S-D link capacity is when $R_1 = R_2$, from which we can solve the optimum value of a as

$$a = \frac{\log_2\left(1 + \frac{g_{rd}P_r}{\eta_d}\right)}{\log_2\left(1 + \frac{g_{rd}P_r}{\eta_d}\right) + \log_2\left(1 + \frac{g_{sr}P_s}{\eta_r}\right)}. \tag{3.45}$$

Alternatively, given a value of a, there is a minimum required transmission power from the source and a minimum required transmission power from the relay in order to achieve a certain S-D link capacity. The minimum transmission power can be calculated from (3.43) and (3.44), respectively.

Compared to $R_{AF,1}$, interference and noise at the relay node has a different effect on $R_{DF,1}$. When $a = 1/2$ and $P_r \to \infty$, (3.43) and (3.44) are the two capacity upper bounds of using AF (see (3.9) and (3.10)). Therefore, with the same transmission power and link conditions, using DF always achieves higher capacity than using AF, because the relay node using DF does not propagate its own experienced interference and noise to the next hop. In a practical system, however, one main advantage of using AF is that the relay node does not need to decode the received signal, which simplifies the implementation at the relay node. Furthermore, using AF makes the two-hop transmissions possible when the link condition between the source and the relay is poor and/or the source transmission power is below the level for the relay node to correctly decode the received signal.

Note also that when using DF, the minimum transmission power levels for P_s and P_r in order to achieve a certain S-D link capacity are independent of each other. Since both hops should decode their received signals, P_s and P_r are not related to each other. In contrast, when using AF, P_s and P_r can trade off between each other. WIthin a certain range, the same S-D link capacity may be achieved by increasing P_s and decreasing P_r or by decreasing P_s and increasing P_r.

Regenerative DF With Direct Link

When the direct link between the source and destination exists, the destination node can receive from the source node in the first phase and from the relay node in the second phase. The destination node then can combine the two copies of the signals. Depending on the codebook used at the relay node for re-encoding the message, there are regenerative and non-regenerative DF. For regenerative DF, the relay decodes the message, re-encodes it using the same codebook as it is used by the source node, and transmits the codeword to the destination. In this case, the two pieces of the signals that the destination receives from the source in the first phase and from the relay node in the second phase are combined together before decoding. The S-D link capacity, R_{RDF}, is given by [6]

$$R_{RDF} = \min\{R_1, R_2\}, \tag{3.46}$$

$$R_1 = a \log_2\left(1 + \frac{g_{sr}P_s}{\eta_r}\right), \tag{3.47}$$

$$R_2 = (1-a)\log_2\left(1 + \frac{g_{sd}P_s}{\eta_d} + \frac{g_{rd}P_r}{\eta_d}\right), \tag{3.48}$$

where R_1 represents the maximum rate at which the relay can reliably decode the source message, and R_2 represents the maximum rate at which the destination can reliably decode the source message given repeated transmissions from the source and the relay nodes.

In order to achieve a certain transmission rate R from the source to the destination, the minimum transmission power required from the source can be solved from (3.47) by letting $R_1 = R$ as

$$P_s \geq \frac{(2^{R/a} - 1)\eta_r}{g_{sr}} \triangleq P_{s,\min}, \tag{3.49}$$

and the minimum required transmission power from the relay node can be solved from (3.48) by letting $R_2 = R$ as

$$P_r = \frac{\eta_d\left[2^{R/(1-a)} - 1 - \frac{P_s g_{sd}}{\eta_d}\right]}{g_{rd}} \triangleq P_{r,\min}. \tag{3.50}$$

From these results we can see that the minimum required P_s does not depend on P_r, while minimum required P_r depends on P_s. Given that $P_s > P_{s,\min}$, larger P_s allows smaller P_r in order to achieve the same S-D capacity.

Non-regenerative DF With Direct Link

For non-regenerative DF, the relay decodes the message, re-encodes it using a different codebook from that used by the source node, and transmits the codeword to the destination. Although the destination receives from the source in the first phase and from the relay in the second phase, the two pieces of the signals are decoded separately. As a result, the S-D link capacity is given by [7]

$$R_{NDF} = \min\{R_1, R_2\}, \tag{3.51}$$

$$R_1 = a \log_2 \left(1 + \frac{g_{sr} P_s}{\eta_r} \right), \tag{3.52}$$

$$R_2 = a \log_2 \left(1 + \frac{g_{sd} P_s}{\eta_d} \right) + (1 - a) \log_2 \left(1 + \frac{g_{rd} P_r}{\eta_d} \right). \tag{3.53}$$

In both the regenerative and non-regenerative DF, the existence of the direct link affects the capacity in both phases of the data transmissions. Increasing P_s can always increase the S-D link capacity, while the transmission of the relay node only affects the capacity in the second phase and has limited effect on the S-D capacity. The effect of P_s and P_r on the S-D link capacity in non-regenerative DF is similar to that in regenerative DF.

For both AF and DF, when the direct link exists, the destination can take advantage of the diversity gain of both the direct and relayed transmissions. This type of communications is referred to as cooperative communications, or cooperative relaying. An application of cooperative communications in wireless cellular networks will be introduced in Section 4.2.

For both non-regenerative and regenerative DF, the link gains of the relay node to the end nodes affect the S-D capacity. Take non-regenerative DF for example, when $g_{sr} \ll g_{sd}$, $R_{NDF} = R_1 \ll R_2$. The S-D capacity is then limited by the poor source-to-relay link conditions. In this case, a straightforward approach to improving the S-D capacity is to have the source transmit in both the time intervals. This motivates the idea in [6] to select the source or the relay to transmit in the second phase based on transmission conditions in the first phase. Similarly, when the relay-to-destination link condition is poor, the second term on the right-hand side of (3.53) has little effect on R_2 as well as the S-D capacity. Having the source transmit in both phases achieves higher capacity than the original non-regenerative DF. The regenerative DF has the same properties.

Source Transmits in Both Phases

Instead of measuring the link gains between the relay and the two end nodes in order to decide which node (source or relay) should transmit in the second phase, one can simply have both the source and the relay nodes transmit in the second phase. The time lines of the nodes in this case are illustrated in Fig. 3.6. In the first phase, the source transmits, and both the relay and the destination receive; in the second phase, both the source and the relay transmit, and the destination receives. We use $P_{s,1}$ and $P_{s,2}$, respectively, to represent the transmission power of the source node in the first and second phases. The S-D link capacity is given by

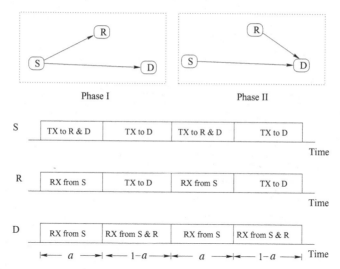

Fig. 3.6 Source transmits in both phases

$$R_{DF,3} = \min\{R_1, R_2\}, \tag{3.54}$$

$$R_1 = a\log_2\left(1 + \frac{P_{s,1}g_{sr}}{\eta_r}\right) + (1-a)\log_2\left(1 + \frac{P_{s,2}g_{sd}}{\eta_d}\right), \tag{3.55}$$

$$R_2 = a\log_2\left(1 + \frac{P_{s,1}g_{sd}}{\eta_d}\right) + (1-a)\log_2\left(1 + \frac{P_{s,2}g_{sd}}{\eta_d} + \frac{P_r g_{rd}}{\eta_d}\right). \tag{3.56}$$

In (3.55), the first term on the right-hand side represents the maximum achievable rate in the source-to-relay transmissions in the first phase. The same rate can be passed to the destination in the second phase, if the achievable rate of the relay-to-destination link in the second phase is no less than this. Together with the achievable rate in the direct link from the source to the destination in the second phase, which is the second term on the right-hand side of (3.55), R_1 gives the upper limit of the achievable rate between the source and the destination. For R_2, the first term on the right-hand side of (3.56) is the achievable rate of the direct link from the source to the destination in the first phase, and the second term is the achievable rate of the transmissions in the second phase, when the destination combines the signals from both the source and the relay. Detailed derivation of $R_{DF,3}$ can be found from [8].

Fig. 3.7 shows the S-D link capacity for different implementations of DF when the direct link between the source and the destination exists. It is seen that having the source node transmit in both phases achieves the highest capacity among the three cases, and using non-regenerate DF in general achieves higher capacity than using regenerative DF. In addition, the capacity difference between using non-regenerative and regenerative DF depends on specific conditions: the S-D capacity is the same in both non-regenerative and regenerative DF when the capacity in each case is limited by the source-to-relay hop; otherwise, the capacity difference between the two cases increases as the relay-to-destination transmission conditions become better.

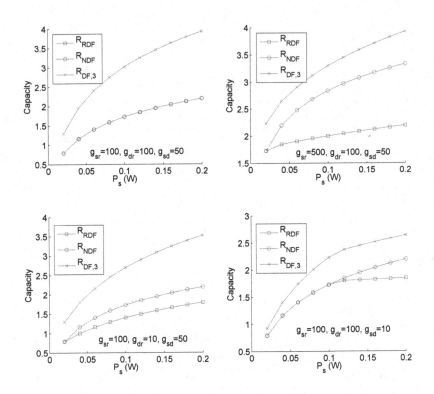

Fig. 3.7 S-D link capacity vs. source transmission power ($P_r = 0.1$W, $\eta_r = \eta_d = \eta$, and link gains are normalized to η.)

3.3.2 Two-Way Relaying

When the data transmissions are in both directions, i.e., the source has data to be transmitted to the destination, and the destination also has data to be transmitted to the source, using relaying with DF requires three time slots for the two end nodes to exchange one pair of packets, one in each direction. In the first time slot, the source node transmits packet X_1 to the relay node; and in the second time slot, the destination node transmits packet X_2 to the relay node. After decoding X_1 and X_2 in the first two time slots, the relay node transmits $X_1 \oplus X_2$ to both the source and the destination nodes in the third time slot. The decoding at the two end nodes in the third time slot can exploit the fact that each of them already knows its prior transmitted signal in the first two time slots. Each of the end nodes decodes the packet from the other given the side information on its own prior transmitted packet.

For X_1, the received power at the relay node is $P_s g_{sr}$; and for X_2, the received power at the relay node is $P_d g_{dr}$. The capacity for each of the source-to-relay link

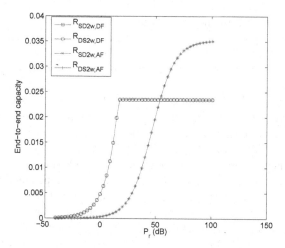

Fig. 3.8 Capacity for two-way relaying using DF and AF ($P_s/g_{sr} = P_d/g_{dr} = 16\text{dB}$, $\eta_s = \eta_d = \eta_r = \eta$, and all link gains are normalized to η)

and the destination-to-relay link in the first two time slots is given by

$$R_{SR} = \frac{1}{3}\log_2\left(1 + \frac{P_s g_{sr}}{\eta_r}\right),\tag{3.57}$$

$$R_{DR} = \frac{1}{3}\log_2\left(1 + \frac{P_d g_{dr}}{\eta_r}\right),\tag{3.58}$$

and the capacity for each link in the third time slot is given by

$$R_{RS} = \frac{1}{3}\log_2\left(1 + \frac{P_r g_{rs}}{\eta_s}\right),\tag{3.59}$$

$$R_{RD} = \frac{1}{3}\log_2\left(1 + \frac{P_r g_{rd}}{\eta_d}\right),\tag{3.60}$$

where the factor $1/3$ is due to the fact that each node only transmits one third of the time. Overall, the S-D link capacity and the D-S link capacity, respectively, are given by

$$R_{SD2w,DF} = \min\{R_{SR}, R_{RD}\},\tag{3.61}$$

$$R_{DS2w,DF} = \min\{R_{DR}, R_{RS}\}.\tag{3.62}$$

Fig. 3.8 compares the link capacity for bidirectional communications using DF and AF, from which we can see that when P_r is relatively small, using DF achieves higher capacity; and when P_r is large, using AF achieves higher capacity. This is an interesting phenomenon. In the two-way relaying scenario, each bi-directional communication between the end nodes requires two time slots for AF and three time slots for DF. Therefore, the capacity using AF can be 1.5 times of that using DF in high SINR region. On the other hand, the fact that interference and noise at the relay node is amplified and propagated to the end nodes negatively affects the performance of AF, which results in significant throughput reduction for AF in low SINR region.

The two-way relaying of using DF introduced above requires one extra time slot than using AF. This is to guarantee time orthogonal transmissions between the source-to-relay and the destination-to-relay transmissions, so that the relay node can decode the received signals from both the end nodes. Instead of having the source and destination nodes transmit at two different time slots, they can transmit concurrently in the first time slot, and their signals contain independent messages for each other to the relay node. However, in order to decode both the signals, the relay node should apply multiuser detection. After correctly decoding the signals from both the source and the destination nodes, the relay node combines them and broadcasts to both the end nodes. In this way, two time slots are required for the two end nodes to exchange one pair of packets using DF. Further analysis about this type of transmissions is available in [9]-[11].

References

1. Cai J, Shen X, Mark JW, Alfa AS (2008) Semi-distributed user relaying algorithm for amplify-and-forward wireless relay networks. IEEE Transactions on Wireless Communications 7(4): 1358-1347.
2. Awad M, Shen X, Zogheib B (2009) Ergodic mutual information of OFDMA-based selection-decode-and-forward cooperative relay networks with imperfect CSI. Physical Communication (Elsevier) 2(3): 184-193.
3. Phan KT, Le-Ngoc T, Vorobyov SA, Tellambura C (2009) Power allocation in wireless multi-user relay networks. IEEE Transactions on Wireless Communications 8(5): 2535-2545.
4. Zhao Y, Adve R, Lim TJ (2007) Improving amplify-and-forward relay neworks: optimal power allocation versus selection. IEEE Transactions on Wireless Communications 6(8): 3114-3123.
5. Rankov B, Wittneben A (2007) Spectral efficient protocols for half-duplex fading relay channels. IEEE Journal on Selected Areas in Communications 25(2): 379-389.
6. Laneman JN, Tse DNC, Wornell GW (2004) Cooperative diversity in wireless networks: efficient protocols and outage behavior. IEEE Transactions on Information Theory 50(12): 3062-3080.
7. Maric I, Yates RD (2010) Bandwidth and power allocation for cooperative strategies in Gaussian relay networks. IEEE Transactions on Information Theory 56(4): 1880-1889.
8. Host-Madsen A, Zhang J (2005) Capacity bounds and power allocation for wireless relay channels. IEEE Transactions on Information Theory 51(6): 2020-2040.
9. Oechtering TJ, Schnurr C, Bjelakovic I, Boche H (2008) Broadcast capacity region of two-phase bidirectional relaying. IEEE Transactions on Information Theory 54(1): 454-458.
10. Xie LL (2007) Network coding and random binning for multi-user channels. Proc. 10th Canadian Workshop on Information Theory: 85-88.
11. Kim SJ, Mitran P, Tarokh V (2007) Performance bounds for bi-directional coded cooperation protocols. 27th International Conference on Distributed Computing Systems Workshops: 83-89.

Chapter 4
Advanced Wireless Communication Networks

Abstract In a conventional wireless cellular communication network, mobile stations (MSs) directly communicate with their associated base stations (BSs), which are the capacity bottleneck and limit the amount of traffic that can be supported in the system. The maximum transmission power of the stations limits the BS coverage. As new techniques are developed, MSs can have multiple air interfaces and communicate with each other either through in-band or out-of-band channels. Appropriately making use of the peer-to-peer communications may increase cellular network capacity, improve quality of service (QoS) to the users, and reduce communication costs. In this chapter we introduce several advanced networks that apply MS-to-MS communications in wireless cellular networks. In Sections 4.1 and 4.2, MSs not having their own traffic can relay traffic for other MSs. In Section 4.1 the MS-to-MS relaying uses out-of-band channel, while in Section 4.2 in-band and cooperative relaying is used for peer MSs to relay traffic for each other. In Section 4.3, MSs that communicate directly with each other form a cognitive radio network, which coexists with the cellular network and utilizes the radio resources of the cellular network through spectrum underlay. Power allocations are studied for each of these networks, and the network performance is analyzed.

Keywords: Out-of-band relay, cooperative relay, in-band relay, cooperative communications, cognitive radio network, power distribution, quality of service.

4.1 Out-of-band Relaying in Wireless Cellular Networks

4.1.1 Relaying Description

The fact that all MSs must directly communicate with the BSs limits the performance of the conventional wireless cellular networks. For the uplink transmissions, MSs closer to the boundary of the BS coverage or in deep channel fading should transmit much higher power than other MSs. In addition to draining the battery power of the MSs significantly, the high transmission power also causes high interference to other users, reduces the service quality of their transmissions, and decreases the system capacity. Furthermore, the limited maximum transmission power may prevent the MSs with poor channel conditions from being connected to the BS. In the downlink, MSs with poor channel conditions to the BS require high transmission power from the BSs in order to receive the same service quality as other MSs, causing unfair resource usage among the users. The poor channel conditions also reduce the downlink capacity. When the traffic load in a certain local area is significantly higher than that in surrounding areas, users in the hotspot experience much worse QoS than in other areas. The traditional approach to balancing traffic loads among the BSs is to force MSs in the heavily loaded BSs to handoff to neighboring lightly loaded BSs, but this does not help if the hotspot is in the coverage area of one BS but not of any other BSs. Using multihop relaying can be an effective approach to shedding traffic from the hotspot to other BSs, and help MSs in the hotspot receive better services.

The multihop relaying can happen in different ways. In the first option, relaying infrastructure is pre-installed, such as the system in [1]. Alternatively, the relaying functionality can be incorporated into the MSs themselves, so that the MSs can relay traffic for each other [2]. This latter option is considered in this section and in Section 4.2. The system considered in this section uses out-of-band relaying. That is, the operations of relaying do not consume the capacity of the cellular networks or inject interference to the cellular air interface. In Section 4.2 in-band relay is considered, where the relay transmissions share the same spectrum as the cellular transmissions.

Consider a CDMA-based cellular system as the one in Section 2.5. In the traditional system, the MSs always use their CDMA radio to communicate directly with the BS. Many current mobile devices are multimode, for example, having both macro-cellular and IEEE 802.11 air interfaces [3]. A typical application for such dual-mode handsets is that when they are within range of an appropriate wireless local area network (WLAN) coverage area, the IEEE 802.11 air interface is used, and when the WLAN coverage is unavailable or large coverage is required, the cellular air interface is activated. In addition to lower communication cost, using the WLAN service can achieve higher transmission rate and reduce the MS power consumption. We consider that the MSs are equipped with both the cellular and ad hoc air interfaces. Those MSs that do not have their own traffic are available as RSs for other MSs. An MS can communicate directly to the BS using its CDMA air interface, or via a relay station (RS) using its ad hoc air interface. In the rest of this section, we first look at transmission power allocations of the network in both the uplink and downlink, and then discuss different criteria for selecting RSs. The outage and rate performance of the network using different RS selection criteria is demonstrated at the end.

4.1.2 Power Allocations

We consider that the MSs do not have the functionality to schedule transmissions of multiple connections, i.e., an MS can either carry its own connection or function as an RS for another MS at one time. All traffic still travels through the BSs, i.e., no traffic is removed from or injected into the system by the ad hoc infrastructure. Since the ad hoc system capacity is often much higher than the total CDMA data rate, the achievable system capacity is determined by that of the CDMA air interface, but not that of the ad hoc air interface. The smaller transmission range of the ad hoc air interface further increases its effective capacity due to spatial reuse of the channel. For this reason we will focus our attention on the capacity of the CDMA BS. It should be kept in mind, however, that under different system design assumptions the capacity of the ad hoc air interface could limit the overall capacity of the system, such as the system with in-band relaying studied in the next section.

We consider homogeneous traffic to simplify the capacity analysis. Each connection requires a constant transmission rate R and a target SINR γ^* at the receiver. Assume that perfect power control is achieved in both the uplink and the downlink. That is, for each connection its actual SINR at the CDMA receiver is equal to the target SINR γ^*, so that the system does not waste any power resources. When ad hoc relaying is used, the original one-hop cellular connection is replaced with a two-hop connection. In the uplink, the first hop is from the source MS to the RS, and the second hop is from the RS to the BS. In the downlink, the first hop is from the BS to the RS, and the second hop is from the RS to the destination MS. Communication between the BS and the RS is done using CDMA, and that between the source/destination MS and the RS is through the ad hoc air interface. In many cases, such as non-real-time data transmissions or voice over IP, the BER in the ad hoc link will effectively be zero due to packet ARQ re-transmissions at the link layer. Therefore, the BER requirement and the SINR threshold in order to achieve this BER for a two-hop connection is effectively the same as that of a one-hop connection. The source/destination MS may use direct communication between itself and the BS, if it turns out no suitable RS is available based on the RS selection criterion, which will be discussed later on. The total number of connections associated to BS b, denoted as $N_{h,b}$, includes those connected to the BS directly (one hop connections) and through RSs (two hop connections).

We define effective link gain (ELG), h_{ib}, for connection i (which is the connection carried by source/destination MS i) associated with BS b. For a two-hop connection, the ad hoc link between the RS and the source/destination MS does not affect the system capacity, but the cellular link between the RS and the BS affects the CDMA capacity. The ELG for the connection is defined as the link gain between the RS and the BS. For a one-hop connection i, the ELG is the link gain between the source/destination MS and the BS. Let g_{ib} represent the link gain between BS b and MS i. The ELG can be defined as

$$h_{ib} = \begin{cases} g_{ib}, & \text{for a one hop connection,} \\ g_{m_i^*b}, & \text{for a two hop connection,} \end{cases} \tag{4.1}$$

where m_i^* is the RS for connection i.

For the uplink, all the connections associated to the same BS should have the same target receiving power at the BS. Define $S_{h,b}$ as the target receiving power for each connection associated to BS b, and a column vector $\mathbf{S}_h = (S_{h,1}, S_{h,2}, \ldots, S_{h,B})^T$. Following the analysis in Section 2.5, a formula similar to (2.28) can be obtained as follows to find the target receiving power,

$$(\mathbf{I} - \boldsymbol{\Delta}_{hu}\mathbf{H}_u)\mathbf{S}_h = \boldsymbol{\eta}_{hu}, \tag{4.2}$$

where

$$\boldsymbol{\Delta}_{hu} = \mathrm{diag}\left(\frac{1}{N_p - N_{h,1}}, \frac{1}{N_p - N_{h,2}}, \ldots, \frac{1}{N_p - N_{h,B}}\right), \tag{4.3}$$

$\mathbf{H}_u$ is a $B \times B$ matrix, whose bth row and cth column is given by

$$H_{u,bc} = \begin{cases} 0, & \text{if } b = c \\ \sum_{i \in \mathscr{B}_c} \frac{h_{ib}}{h_{ic}}, & \text{if } b \neq c \end{cases} \tag{4.4}$$

and $\boldsymbol{\eta}_{hu}$ is a column vector, whose bth element is given by

$$\eta_{hu,b} = \frac{\eta_b}{N_p - N_{h,b}}. \tag{4.5}$$

For the downlink, let $P_{h,c}$ be the total transmission power of BS c, and $\mathbf{P}_h = (P_{h,1}, P_{h,2}, \ldots, P_{h,B})^T$. We have

$$(\mathbf{I} - \boldsymbol{\Delta}_{hd}\mathbf{H}_d)\mathbf{P}_h = \boldsymbol{\eta}_{hd}, \tag{4.6}$$

where

$$\boldsymbol{\Delta}_{hd} = \mathrm{diag}\left(\frac{1}{\xi(N_{\xi,p} - N_{h,1})}, \frac{1}{\xi(N_{\xi,p} - N_{h,2})}, \ldots, \frac{1}{\xi(N_{\xi,p} - N_{h,B})}\right), \tag{4.7}$$

$\mathbf{H}_d$ is a $B \times B$ matrix, whose cth row and bth column element is given by

$$H_{d,cb} = \begin{cases} 0, & \text{if } c = b \\ \sum_{i \in \mathscr{B}_c} \frac{h_{ib}}{h_{ic}}, & \text{if } c \neq b \end{cases} \tag{4.8}$$

and $\boldsymbol{\eta}_{hd}$ as a column vector with the cth element given by

$$\eta_{hd,c} = \frac{\sum_{i \in \mathscr{B}_c} \frac{\eta_i}{h_{ic}}}{\xi(N_{\xi,p} - N_{h,c})}. \tag{4.9}$$

Based on the analysis in Section 2.5, a feasible solution exists to (4.2) if $\rho(\boldsymbol{\Delta}_{hu}\mathbf{H}_u) < 1$, and a feasible solution exists to (4.6) if $\rho(\boldsymbol{\Delta}_{hd}\mathbf{H}_d) < 1$. By looking at the matrices $\mathbf{H}_u$ and $\mathbf{H}_d$ we find that the non-diagonal elements of the matrices are summations of normalized ELGs, each of which is the ratio of the ELG from an interfering connection to the ELG of a given connection. Further looking at the diagonal matrices Δ_{hu} and Δ_{ud} we find that each diagonal element of the matrices is the inverse of the difference between the pole capacity of the cell and the actual number of connections associated to the cell. Based on the analysis in Section 2.5, in order to reduce the possibility of infeasible power distributions and increase the system capacity, one approach is to reduce the normalized ELGs, and the other approach is to reduce the number of connections associated in the cell with large $N_{h,b}$ (close to or larger than the pole capacity of a single cell). Both objectives may be achieved through appropriately choosing the RSs.

By comparing (4.4) and (4.8) we can see that $\mathbf{H}_u$ is the transpose of $\mathbf{H}_d$. Using RSs that can reduce $\sum_{i \in \mathscr{B}_c} \frac{h_{ib}}{h_{ic}}$ reduces both the bth row and cth column in $\mathbf{H}_u$ and the cth row and bth column in $\mathbf{H}_d$ ($b \neq c$). By comparing (4.3) and (4.7) and comparing (4.5) and (4.9) we can see that if cell c is a hotspot cell, relaying traffic from cell c to a neighboring under-loaded cell b can reduce both $\Delta_{hu,c}$ and $\eta_{hu,c}$ for the uplink and $\Delta_{hd,c}$ and $\eta_{hd,c}$ for the downlink. These observations indicate that if an RS can improve the uplink capacity, it should also improve the downlink capacity. For this reason we do not have to make separate criteria for the uplink and the downlink when selecting the RSs. Below we describe the RS selection criteria for the uplink.

4.1.3 Relay Station Selection

Based on the above analysis, the ad hoc relaying may be achieved in two basic approaches. The first is to relay traffic within the same cell, and the second is to relay traffic from the hotspot cells to the neighboring lightly loaded cells. For relaying traffic within the same cell, the use of ad hoc relaying is to adjust h_{ib}/h_{ic} values, so that the RS can access the same BS as the source MS but with better link quality. Moving traffic from hotspot cells to neighboring non-hotspot cells can reduce the value of $N_{h,c}$ and decrease the number of summed terms in $\sum_{i \in \mathscr{B}_c} h_{ib}/h_{ic}$ (if cell c is a hotspot cell). Since all the cells share the same bandwidth, an optimum approach to selecting RSs requires global information regarding traffic load in all the cells and the transmission conditions of all the connections. This is difficult to implement in a practical system. Reference [4] proposed several simplified RS selection criteria, which are described below. Before introducing these criteria, we define $\mathscr{A}_i$ as a set of the potential RSs available for connection i.

Relaying traffic within the same cell

The first RS selection criterion is based on the link gains. When selecting an RS for connection i in cell c, among all the potential RSs, the one having the best link gain to BS c is selected as the RS. That is, MS m_i^* is selected as the RS, if

$$m_i^* = \arg \max_{m \in \mathscr{A}_i} g_{mc}. \tag{4.10}$$

Instead of using the link gains, the second RS selection criterion is based on distances. Among all the available RSs, the one closest to the BS is selected as the RS. That is,

$$m_i^* = \arg \min_{m \in \mathscr{A}_i} d_{mc}, \qquad (4.11)$$

where d_{mc} is the distance between MS m and BS c.

The third criterion is to choose the available RS based on the normalized ELGs. As the normalized ELG integrates both the transmission and interference conditions of a link, using this criterion is more comprehensive, compared to the previous two criteria based on link gains or distances. We notice that each non-diagonal element in the matrix $\mathbf{H}_u$ is a summation of normalized ELGs, each of which is a ratio of the ELG of an interfering link to the ELG of the desired link. Since the impact of the inter-cell interference is most significant in the immediate neighboring cells and (its mean value) decays geometrically with distance, we can simplify the RS selection for a connection by considering only the interference to and from the immediate neighboring cells. In this way, global information about the ELGs of the connections is not necessary, but only that to their neighboring BSs is required. As is normally the case in CDMA networks, all active MSs choose their BS by monitoring pilot signals transmitted by neighboring BSs. The link gains to the neighboring BSs are available at each MS for the purpose of searching the serving BS and assisting handoffs. The criterion for RS selection is to choose MS m_i^* as the RS for connection i, $i \in \mathscr{B}_c$, according to the following condition:

$$m_i^* = \arg \min_{m \in \mathscr{A}_i} \max_{b \in \mathscr{N}_c} \frac{g_{mb}}{g_{mc}}, \qquad (4.12)$$

where $\mathscr{N}_c$ represents a set of neighboring cells of cell c. Intuitively, a smaller h_{mc} represents a poor communication link between the RS and the serving BS, and a larger g_{mb} means that transmissions of the RS can cause high interference to cell b.

Shedding traffic from hotspot cells

When some cells have much higher traffic load than other cells, another way to increase the chance for feasible power distribution is to decrease the number of active connections in the hotspot cells. If a hotspot is located in the coverage area of BS c, this method decreases $N_{h,c}$ and the number of summed terms in $H_{u,bc}$ for the uplink and $H_{d,cb}$ for the downlink ($b \neq c$). The traffic load in a hotspot cell can be close to or even higher than the pole capacity if relay is not used. Moving connections from hotspot cell c to a neighboring non-hotspot cell b increases $N_{h,b}$. Therefore, selecting the BS to which a connection is moved is important. A simple rule is to choose the neighboring BS with the smallest number of associated connections. The RS selection criterion can be the same as for relaying traffic within the same cell, except that the serving BS is different. Let BS c^* represent the BS that connection i is shed into, and m_i^* represent the selected RS. The RS selection criteria based on link gains, distances, and ELGs, respectively, are given by

$$m_i^* = \arg\max_{m \in \mathscr{A}_i} g_{mc^*}, \tag{4.13}$$

$$m_i^* = \arg\min_{m \in \mathscr{A}_i} d_{mc^*}, \tag{4.14}$$

and

$$m_i^* = \arg\min_{m \in \mathscr{A}_i} \max_{b \in \mathscr{N}_{c^*}} \frac{g_{mb}}{g_{mc^*}}. \tag{4.15}$$

4.1.4 Performance Results

We consider the same system as in Fig. 2.7, and compare the performance of the above RS selection criteria. The same default parameters are used as in Section 2.6. The CDMA cell size is defined as the maximum distance from the BS to the cell boundary, and is normalized to 1. The ad hoc coverage of each MS is circular with radius r, which is normalized with respect to the cell size. The probability that an MS carries its own active traffic is defined as p_a. Therefore, on average a fraction of $1 - p_a$ MSs are available for relaying traffic for their peers. For relaying traffic within the same cell, all three RS selection criteria are considered. For shedding traffic from the hotspot cell to neighboring cells, the criterion based on ELG is used.

Transmission rate performance

Assuming all the connections transmit at the same rate, we first look at the maximum transmission rate of each connection, which is obtained using the same algorithm as in Section 2.6.

We first consider that all MSs are uniformly distributed throughout the system coverage area. For $p_a = 0.5$ in all the cells, the maximum transmission rate is shown in Figs. 4.1-4.4, where three different RS selection criteria for relaying traffic within the same cell are shown, and "shed from hotspot" represents the ad hoc relaying that sheds traffic from cell 1 to neighboring cells. Since cell 1 (the center cell) experiences higher interference than other cells, it is "hotter" compared to other cells. For comparison, the rates in a system using the hard and soft handoff without ad hoc relaying are also shown, and the criteria for choosing serving BSs in these cases are the same as in Section 2.6. Figs. 4.1 and 4.2 show that as the number of MSs increases, the achieved transmission rate decreases for all the RS selection criteria. Among all, the criterion based on the ELGs achieves the highest rate in both the uplink and the downlink. All the other RS selection criteria achieve lower rates than using SHO without ad hoc relaying. Using the RS selection criteria based on link gains and distances achieve approximately the same rate as using hard handoff without ad hoc relaying. Shedding traffic from cell 1 to neighboring cells achieves slightly higher rate than relaying within the same cell and using the RS selection criterion based on link-gains or distances, and better than using hard handoff without

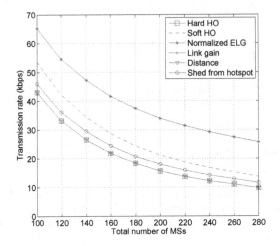

Fig. 4.1 Uplink: average transmission rate vs. number of MSs

relaying. From Figs. 4.1 and 4.2, the advantage of using ad hoc relaying by choosing RSs based on the relative ELGs is obvious.

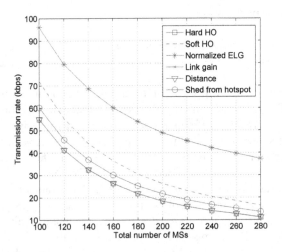

Fig. 4.2 Downlink: average transmission rate vs. number of MSs

The ad hoc coverage limits the number of available RSs and therefore affects the performance improvement of relaying. Figs. 4.3 and 4.4 further show the achievable rate as the ad hoc distance changes. As the ad hoc coverage distance increases, more RSs are available within the coverage area of an active MS. This provides chances for finding better RSs. For this reason, the transmission rate increases as the ad hoc distance increases. Using the RS selection criterion based on the relative ELGs, the transmission rate increases greatly with the ad hoc distance, and can be a lot higher than using soft handoff without relaying. When using the RS selection criteria based on link gains or distances, longer ad hoc distance also results in higher transmission rate, but the rate does not increase with the ad hoc distance very significantly.

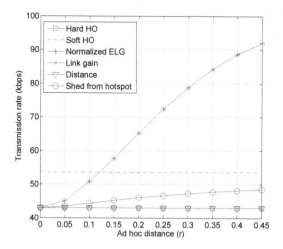

Fig. 4.3 Uplink: average transmission rate vs. ad hoc distance

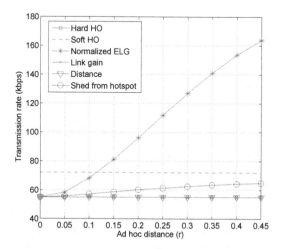

Fig. 4.4 Downlink: average transmission rate vs. ad hoc distance

Outage performance

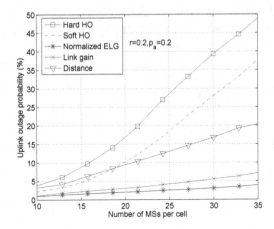

Fig. 4.5 Uplink outage probability vs. number of MSs, uniformly distributed traffic

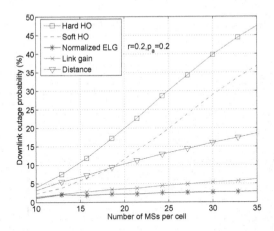

Fig. 4.6 Downlink outage probability vs. number of MSs, uniformly distributed traffic

Next we consider that all the links transmit at a constant rate, and compare the outage performance. When there is no feasible solution to the power allocations in a cell, i.e., the calculated target receiving power in the uplink or transmission power in the downlink are negative, we consider that there is one outage in that cell. In order to reduce the edge effect, simulation results are collected and averaged over the center cell and the six first tier cells. In Figs. 4.5-4.8 we first consider the case when all MSs are uniformly distributed throughout the system coverage area. In Figs. 4.9 and 4.10 we consider the outage performance in a system with a hotspot cell.

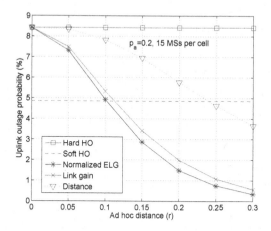

Fig. 4.7 Uplink outage probability vs. ad hoc distance, uniformly distributed traffic

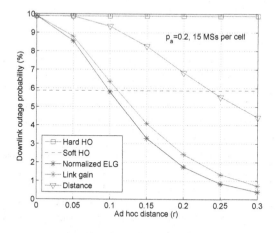

Fig. 4.8 Downlink outage probability vs. ad hoc distance, uniformly distributed traffic

Figs. 4.5 and 4.6 show the outage performance of the RS selection criteria based on link gains, distances, and ELGs. It can be seen that all the three RS selection criteria can improve the outage probability in both the uplink and the downlink,

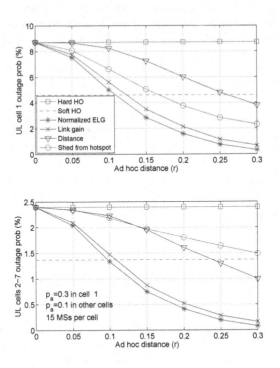

Fig. 4.9 Uplink outage probability vs. ad hoc distance, hotspot case

compared to both the hard and soft handoff without relaying. This effect can be
translated into improved capacity for a given outage probability requirement. For
example, if the maximum tolerable outage probability is 5%, about 11 MSs (in the
uplink and downlink) can be accommodated using the hard handoff, 15 using the
soft handoff, 28 using ad hoc relaying and the RS selection criterion based on link
gains, and more than 35 using ad hoc relaying and the RS selection criterion based
on ELGs. Using the RS selection criterion based on the link gains, an active MS
may find an RS with good link quality to its serving BS, which reduces the required
power and the co-channel interference to other MSs. Using the criterion based on
ELGs performs better than the criterion based on link gains, since the former takes
into consideration both the channel condition to the serving BS and interference
to neighboring BSs. For a given active probability p_a, the outage probability using
the criteria based on ELGs or link gains is relatively insensitive to the traffic load
changes in the system. As the number of MSs increases, traffic load increases, and
the number of available RSs also increases, which provides the active MSs with a
better chance to find RSs with good link quality to the BS. From Figs. 4.5 and 4.6 we
can see that the criterion based on ELGs is less sensitive to traffic load increase than
the criterion based on link gains, and can reduce the outage probability by about

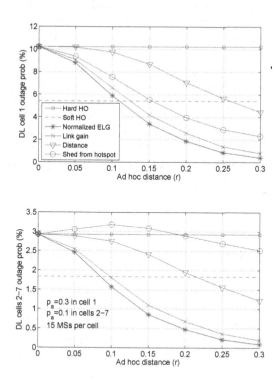

Fig. 4.10 Downlink outage probability vs. ad hoc distance, hotspot case

50% of that based on link gains at high traffic loading. The outage performance of the distance-based criterion is worse than the other two criteria, because it does not consider the actual link gains and interference conditions.

Figs. 4.7 and 4.8 show the outage probability as the ad hoc distance changes, where we can see that the system using ad hoc relaying with any of the three RS selection criteria requires a minimum ad hoc distance in order to achieve lower outage probability than the system without relaying. Using the RS selection criterion based on ELGs requires the shortest ad hoc distance, which is approximately the same as that based on link gains, while using the distance-based RS selection criterion requires much longer ad hoc distance.

When the traffic load is not uniformly distributed, Figs. 4.9 and 4.10 show the performance with the ad hoc relaying performed in the hotspot cell only. In the system considered, all MSs are uniformly distributed within the service area. The probability of being active for each MS in cell 1 is $p_a = 0.3$ and for each MS in other cells is $p_a = 0.1$. In this case, cell 1 is a hotspot cell in conventional cellular communications. When shedding traffic from the hotspot cell, the RSs are selected based on the ELGs. The figures show that by moving traffic from the hotspot cell to neighboring cells, outage probability in the hotspot cell can be significantly reduced in both the uplink and the downlink. Shedding traffic from the hotspot cell has some contradictory effects on the performance in neighboring cells. First, since this increases the traffic load in the neighboring cells, the outage probability in these cells may be increased. On the other hand, as the traffic load in the hotspot cell is reduced, it causes less interference to the neighboring cells, and this may improve the outage performance in the neighboring cells. Overall, the traffic load becomes more balanced in all cells, and the outage probability in all cells can be reduced in most cases, compared to the system without relaying. This is shown in Figs. 4.9 and 4.10. The figures also show that moving traffic inside each cell by choosing RSs based on ELGs is still the best choice for improving the outage performance even in this system.

4.2 Cooperative Relaying in Wireless Cellular Networks

In this section we consider a CDMA-based cellular network with in-band relaying. In traditional in-band relay, the direct transmission between the MS and the BS is replaced with a two hop link, one hop between the MS and the RS, and another hop between the RS and the BS. When such relaying is allowed, MS power consumption can be reduced because of shorter transmission distances. With the reduction in transmission power, co-channel interference may also be reduced, which leads to increased system capacity. Therefore, using relaying in a cellular network can potentially increase the system capacity. On the other hand, such capacity improvement is limited, since the in-band relaying traffic creates its own co-channel interference [5] [6]. Using cooperative relaying may better utilize the radio resources, so that the destination can exploit the diversity provided by both the direct link and the relay link. In this section we study the transmission power allocations in such a network by combining the work from references [7] and [8].

We consider a single cell network, where cooperative relaying is used in both the uplink and the downlink. In the uplink, MSs carrying their own connections are the source stations, and the BS is the destination for all connections. In the downlink, the BS is the source for all connections and the destinations are the MSs carrying their own active traffic. MSs that do not carry active traffic of their own are available for cooperatively relaying packets for their peer stations. The RS can use either decode-and-forward (DF) or amplify-and-forward (AF) techniques (see Chapter 3 for details). For a simple presentation, below we use "source/destination MS" to denote an MS carrying its own active traffic, and use "RS" to denote an MS working as the RS. We consider a synchronized case as follows. For the uplink, all the source MSs transmit in the odd time slots, when the RSs and the BS listens. In the even time slots, the RSs transmit and the BS listens. This is shown in Fig. 4.11. For the downlink, the BS transmits and the MSs and RSs listen in the odd time slots, and the RSs transmit to the MSs in the even time slots. This is shown in Fig. 4.12.

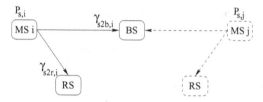

Time slot 1: MSs transmit

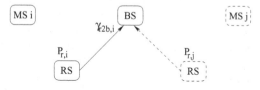

Time slot 2: RSs transmit

Fig. 4.11 Cooperative relaying in the uplink

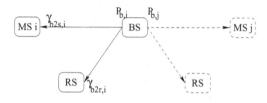

Time slot 1: BS transmits

Time slot 2: RSs transmit

Fig. 4.12 Cooperative relaying in the downlink

At the end of each even time slot, the destination combines the signals from the source and the RS using maximum ratio combining (MRC). Below we study the transmission power allocations in a single cell network.

4.2.1 Power Distribution in the Uplink

We consider N connections in the uplink. The source MS carrying the ith connection in the uplink is denoted as the ith MS. We use $P_{s,i}$ to denote the transmission power of the source MS for connection i, and $P_{r,i}$ the transmission power of the RS for connection i. Let $\gamma_{s2b,i}$ represent the SINR of the received signal at the BS receiver input for a packet transmitted from the source MS, $\gamma_{s2r,i}$ the SINR of the received signal at the RS receiver input for the same packet, and $\gamma_{r2b,i}$ the SINR at the BS receiver input for a packet forwarded from the RS. These SINR notations are indicated in Fig. 4.11. At the end of each even time slot, the SINR of the combined signal at the BS receiver is $\gamma_i = \gamma_{s2b,i} + \gamma_{r2b,i}$ for connection i. In order to ensure successful transmission, $\gamma_i \geq \gamma^*$, where γ^* is the minimum required SINR for correct transmissions.

Consider connection i. In the odd time slots, the RS and the BS receive desired signals from transmissions of their own source MS as well as interference from other MSs. The received SINR for a packet transmitted from MS i at the BS is given by

$$\gamma_{s2b,i} = \frac{W}{R} \frac{P_{s,i} g_{s2b,i}}{\sum_{j=1}^{N} P_{s,j} g_{s2b,j} - P_{s,i} g_{s2b,i} + \eta}, \tag{4.16}$$

for $i = 1, 2, \ldots, N$, where R is the transmission rate, $g_{s2b,i}$ is the link gain from the source MS of connection i to the BS, and η is the background noise power. Similarly, the received SINR of the same packet at the RS receiver is given by

$$\gamma_{s2r,i} = \frac{W}{R} \frac{P_{s,i} g_{s2r,ii}}{\sum_{j=1}^{N} P_{s,j} g_{s2r,ji} - P_{s,i} g_{s2r,ii} + \eta}, \tag{4.17}$$

where $g_{s2r,ji}$ denotes the link gain from the source MS of connection j to the RS of connection i.

When DF is used at the RSs, $\gamma_{s2r,i}$ should be larger than or equal to γ^* so that the RS can successfully decode the received signal from the source before forwarding it to the BS. When the packet is forwarded from the RS to the BS, it experiences interference from transmissions of all other RSs, and its SINR at the BS receiver is

$$\gamma_{r2b,i} = \frac{W}{R} \frac{P_{r,i} g_{r2b,i}}{\sum_{j=1}^{N} P_{r,j} g_{r2b,j} - P_{r,i} g_{r2b,i} + \eta}, \tag{4.18}$$

where $g_{r2b,i}$ denotes the link gain from the RS of connection i to the BS.

When AF is used, successful decoding is not required at the RS, which simply amplifies the received analog signal together with interference and noise, and forwards to the BS. Therefore, the expression of $\gamma_{r2b,i}$ should take the amplified interference into consideration. Based on (3.5) we have

$$\gamma_{r2b,i} = \frac{W}{R} \frac{\gamma_{s2r,i} \frac{P_{r,i} g_{r2b,i}}{\sum_{j=1}^{N} P_{r,j} g_{r2b,j} - P_{r,i} g_{r2b,i} + \eta}}{\gamma_{s2r,i} + 1 + \frac{P_{r,i} g_{r2b,i}}{\sum_{j=1}^{N} P_{r,j} g_{r2b,j} - P_{r,i} g_{r2b,i} + \eta}}, \tag{4.19}$$

which can be rewritten as

$$\gamma_{r2b,i} = \frac{W}{R} \frac{P_{r,i}g_{r2b,i}\beta_{u,i}}{\sum_{j=1}^{N} P_{r,j}g_{r2b,j} - P_{r,i}g_{r2b,i}\beta_{u,i} + \eta}, \tag{4.20}$$

where $\beta_{u,i} = \frac{\gamma_{s2r,i}}{\gamma_{s2r,i}+1}$.

Given the above analysis, an optimum power distribution problem can be formulated as follows:

$$\text{Problem I: } \min_{\mathbf{P}_s, \mathbf{P}_r} \sum_{i=1}^{N} (P_{s,i} + P_{r,i}) \tag{4.21}$$

$$\text{s.t.} \quad \gamma_{s2b,i} + \gamma_{r2b,i} \geq \gamma^*, \ i = 1, 2, \ldots, N \tag{4.22}$$

$$\gamma_{s2r,i} \geq \gamma^*, \ i = 1, 2, \ldots, N, \text{ for DF only} \tag{4.23}$$

$$0 \leq P_{s,i} \leq P_{\text{MS,max}}, \ i = 1, 2, \ldots, N \tag{4.24}$$

$$0 \leq P_{r,i} \leq P_{\text{MS,max}}, \ i = 1, 2, \ldots, N \tag{4.25}$$

where $\mathbf{P}_s = (P_{s,1}, P_{s,2}, \ldots, P_{s,N})$, $\mathbf{P}_r = (P_{r,1}, P_{r,2}, \ldots, P_{r,N})$, $P_{\text{MS,max}}$ is the maximum transmission power of an MS, $\gamma_{s2b,i}$ and $\gamma_{s2r,i}$, respectively, are given by (4.16) and (4.17), and $\gamma_{r2b,i}$ is given by (4.18) for DF and (4.20) for AF. If there is no feasible solution to the optimization problem, one or more connections should be removed so that a feasible solution can be found for the remaining connections. The removed connections are in outage.

Problem I is non-convex. The left-hand side of the constraints in (4.22) and (4.23) are non-linear functions of $P_{s,i}$'s and $P_{r,i}$'s. With some simple mathematical manipulations we can use the iterative method proposed in [10] to solve the problem. The method is summarized in Appendix B. Let $A_i = P_{s,i}g_{s2b,i}$, $\overline{A}_i = \sum_{j=1,j\neq i}^{N} P_{s,j}g_{s2b,j}$, $B_i = P_{r,i}g_{r2b,i}$, $\overline{B}_i = \sum_{j=1,j\neq i}^{N} P_{r,j}g_{r2b,j}$, and $C = \frac{R\gamma^*}{W}$. Then we have $\gamma_{s2b,i} = \frac{W}{R}\frac{A_i}{\overline{A}_i+\eta}$ and $\gamma_{r2b,i} = \frac{W}{R}\frac{B_i}{\overline{B}_i+\eta}$. Then constraint (4.22) can be rewritten as

$$\frac{A_i}{\overline{A}_i + \eta} + \frac{B_i}{\overline{B}_i + \eta} \geq C, \tag{4.26}$$

which can be further rewritten as

$$\frac{C\overline{A}_i\overline{B}_i + \eta C\overline{A}_i + \eta C\overline{B}_i + \eta^2 C}{A_i\overline{B}_i + B_i\overline{A}_i + \eta A_i + \eta B_i} \leq 1. \tag{4.27}$$

In (4.27) both the numerator and the denominator on the left-hand side are posynomials of $P_{s,i}$'s and $P_{r,i}$'s (See Appendix B for a definition of a posynomial). Replacing (4.22) in Problem I with (4.27), and rewrite (4.23) as $\frac{\gamma^*}{\gamma_{s2r,i}} \leq 1$, we can use the iterative method in Appendix B (Case 2) to solve Problem I.

From the above problem formulation we find that in order to solve the optimization problem, the BS should know link gains from all source MSs to all RSs. This information cannot be easily obtained by the BS. A practical solution is to assign a fixed SINR target to each of the MS-to-BS link and the RS-to-BS link, so that distributed power control is possible. For example, each of these two links can have a fixed share of the total required SINR for the given connection. A simple scheme is to determine the shares based on their link gains to the BS as follows:

$$\frac{\gamma_{s2b,i}}{\gamma_{r2b,i}} = \frac{g_{s2b,i}}{g_{r2b,i}}, \text{ and} \tag{4.28}$$

$$\gamma_{s2b,i} + \gamma_{r2b,i} = \gamma^*, \tag{4.29}$$

for $i = 1,2,\ldots,N$. Based on this policy, the target SINRs for the MS-to-BS link and the RS-to-BS link are $\frac{g_{s2b,i}}{g_{s2b,i}+g_{r2b,i}}\gamma^*$ and $\frac{g_{r2b,i}}{g_{s2b,i}+g_{r2b,i}}\gamma^*$, respectively. When the dynamic power control as introduced in Section 1.2 converges and the power control is perfect, we have

$$\gamma_{s2b,i} = \frac{W}{R} \frac{P_{s,i}g_{s2b,i}}{\sum_{j=1}^{N} P_{s,j}g_{s2b,j} - P_{s,i}g_{s2b,i} + \eta} = \frac{g_{s2b,i}\gamma^*}{g_{s2b,i}+g_{r2b,i}}, \tag{4.30}$$

$$\gamma_{r2b,i} = \frac{W}{R} \frac{P_{r,i}g_{r2b,i}}{\sum_{j=1}^{N} P_{r,j}g_{r2b,j} - P_{r,i}g_{r2b,i} + \eta} = \frac{g_{r2b,i}\gamma^*}{g_{s2b,i}+g_{r2b,i}}, \text{ for DF} \tag{4.31}$$

$$\gamma_{r2b,i} = \frac{W}{R} \frac{P_{r,i}g_{r2b,i}\beta_{u,i}}{\sum_{j=1}^{N} P_{r,j}g_{r2b,j} - P_{r,i}g_{r2b,i}\beta_{u,i} + \eta} = \frac{g_{r2b,i}\gamma^*}{g_{s2b,i}+g_{r2b,i}}, \text{ for AF} \tag{4.32}$$

for $i = 1,2,\ldots,N$. Mathematically, the transmission power of each source MS and RS can be found from equations (4.30) and (4.31) for DF and equations (4.30) and (4.32) for AF. The solution to the above equation system is feasible if all the following conditions are satisfied: i) $0 \le P_{s,i} \le P_{\text{MS,max}}$ for all $i = 1,2,\ldots,N$, and ii) $0 \le P_{r,i} \le P_{\text{MS,max}}$ for all $i = 1,2,\ldots,N$. Note that this power distribution does not guarantee the SINR requirement of the MS-to-RS link for DF. If the solution does not satisfy the condition $\gamma_{s2r,i} \ge \gamma^*$, successful decoding at the RS is not possible for DF. In this case, one connection is removed (and in outage), and the transmission power for the remaining connections is recalculated.

4.2.2 Power Distribution in the Downlink

In the downlink the BS is the source for all connections. Let $P_{b,i}$ denote the transmission power of the BS for connection i, and $P_{r,i}$ denote the transmission power of the RS for connection i. Then $P_b = \sum_{i=1}^{N} P_{b,i}$ is the total transmission power of the BS, and $\sum_{j=1}^{N} P_{r,j}$ is the total transmission power of all RSs. Below we use $\gamma_{b2d,i}$ to represent the SINR of the received signal at the destination MS for a packet transmitted from the BS for connection i, $\gamma_{b2r,i}$ the SINR at the RS for the same packet, and $\gamma_{r2d,i}$ the received SINR at the destination MS for the packet forwarded from the RS. These notations are illustrated in Fig. 4.12. At the end of each even time slot, the destination MS of connection i combines the signal from the BS in the previous time slot and the signal from the RS in the current time slot using MRC, and the SINR of the combined signal is $\gamma_i = \gamma_{b2d,i} + \gamma_{r2d,i}$. The objective of power distribution in the downlink is to guarantee the SINR requirements of the connections, while minimizing $P_b + \sum_{i=1}^{N} P_{r,i}$.

At the odd time slots when the BS transmits, the received SINR of the signal for connection i at the destination receiver is given by

$$\gamma_{b2d,i} = \frac{W}{R} \frac{P_{b,i} g_{b2d,i}}{P_b g_{b2d,i} - P_{b,i} g_{b2d,i} + \eta}, \tag{4.33}$$

where $g_{b2d,i}$ is the link gain from the BS to the destination MS of connection i. The same packet also reaches the RS of connection i with a received SINR given by

$$\gamma_{b2r,i} = \frac{W}{R} \frac{P_{b,i} g_{b2r,i}}{P_b g_{b2r,i} - P_{b,i} g_{b2r,i} + \eta}, \tag{4.34}$$

where $g_{b2r,i}$ is the link gain from the BS to the RS of connection i.

For DF, the RS should decode the received packet, re-encode, and forward to the destination. That is, $\gamma_{b2r,i} \geq \gamma^*$ should hold. At the destination receiver, the forwarded signal from the RS experiences interference from transmissions of all other RSs. Assuming the RS can correctly decode the signal from the BS, the SINR at the destination receiver for connection i is given by

$$\gamma_{r2d,i} = \frac{W}{R} \frac{P_{r,i} g_{r2d,ii}}{\sum_{j=1}^{N} P_{r,j} g_{r2d,ji} - P_{r,i} g_{r2d,ii} + \eta}, \tag{4.35}$$

where $g_{r2d,ji}$ is the link gain from the RS for connection j to the destination MS of connection i.

For AF, the RS does not decode the received signal, but amplifies the received analog signal, interference, and noise ,and forwards to the destination. Similar to the derivation of $\gamma_{r2b,i}$ in the uplink, the SINR of the forwarded signal at the destination can be derived as

$$\gamma_{r2d,i} = \frac{W}{R} \frac{\beta_{d,i} P_{r,i} g_{r2d,ii}}{\sum_{j=1}^{N} P_{r,j} g_{r2d,ji} - \beta_{d,i} P_{r,i} g_{r2d,ii} + \eta}, \tag{4.36}$$

where $\beta_{d,i} = \frac{\gamma_{b2r,i}}{\gamma_{b2r,i}+1}$.

Similar to that for the uplink, the optimum power distribution problem for the downlink is then formulated as follows:

$$\text{Problem II:} \min_{P_b, \mathbf{P}_r} \left(P_b + \sum_{i=1}^{N} P_{r,i} \right) \tag{4.37}$$

$$\text{s.t.} \quad \gamma_{b2d,i} + \gamma_{r2d,i} \geq \gamma^*, \ i = 1,2,\ldots,N \tag{4.38}$$

$$\gamma_{b2r,i} \geq \gamma^*, \ i = 1,2,\ldots,N, \text{ for DF only} \tag{4.39}$$

$$0 \leq P_{r,i} \leq P_{\text{MS,max}}, \ i = 1,2,\ldots,N \tag{4.40}$$

$$0 \leq P_b \leq P_{\text{BS,max}} \tag{4.41}$$

where $\mathbf{P}_r = (P_{r,1}, P_{r,2}, \ldots, P_{r,N})$, $P_{BS,\max}$ is the maximum transmission power of the BS, $\gamma_{b2d,i}$ and $\gamma_{b2r,i}$, respectively, are given by (4.33) and (4.34), and $\gamma_{r2d,i}$ is given by (4.35) for DF and (4.36) for AF. A similar method as in Section 4.2.1 can be used to convert Problem II into a format that can be solved by the iterative method in Appendix B (Case 2). Communication outage occurs if the optimization problem does not have a feasible solution. The optimum solution for the downlink is not practical, since it requires link gains from all the RSs to all the destination MSs.

A similar heuristic power distribution solution as in the uplink can be designed for the downlink. The total required SINR of each downlink connection is split between the BS-to-MS and RS-to-MS links based on their link gains as follows

$$\frac{\gamma_{b2d,i}}{\gamma_{r2d,i}} = \frac{g_{b2d,i}}{g_{r2d,ii}}, \text{ and} \tag{4.42}$$

$$\gamma_{b2d,i} + \gamma_{r2d,i} = \gamma^*, \tag{4.43}$$

for $i = 1, 2, \ldots, N$. Based on this policy, the target SINRs for the BS-to-MS link and the RS-to-MS link are $\frac{g_{b2d,i}}{g_{b2d,i}+g_{r2d,ii}}\gamma^*$ and $\frac{g_{r2d,ii}}{g_{b2d,i}+g_{r2d,ii}}\gamma^*$, respectively. When the dynamic power control converges and the power control is perfect, we have

$$\gamma_{b2d,i} = \frac{W}{R} \frac{P_{b,i}g_{b2d,i}}{P_b g_{b2d,i} - P_{b,i}g_{b2d,i} + \eta} = \frac{g_{b2d,i}\gamma^*}{g_{b2d,i} + g_{r2d,ii}}, \tag{4.44}$$

$$\gamma_{r2d,i} = \frac{W}{R} \frac{P_{r,i}g_{r2d,ii}}{\sum_{j=1}^{N} P_{r,j}g_{r2d,ii} - P_{r,i}g_{r2d,ii} + \eta} = \frac{g_{r2d,ii}\gamma^*}{g_{b2d,i} + g_{r2d,ii}}, \text{ for DF}, \tag{4.45}$$

$$\gamma_{r2d,i} = \frac{W}{R} \frac{\beta_{d,i}P_{r,i}g_{r2d,ii}}{\sum_{j=1}^{N} P_{r,j}g_{r2d,ji} - \beta_{d,i}P_{r,i}g_{r2d,ii} + \eta} = \frac{g_{r2d,ii}\gamma^*}{g_{b2d,i} + g_{r2d,ii}}, \text{ for AF}. \tag{4.46}$$

Mathematically, the transmission power of the BS and RS for each connection can be solved using equations (4.44) and (4.45) for DF and using equations (4.44) and (4.46) for AF. The solution is feasible if i) $0 \leq \sum_{i=1}^{N} P_{b,i} \leq P_{BS,\max}$, and ii) $0 \leq P_{r,i} \leq P_{MS,\max}$ for $i = 1, 2, \ldots, N$. For DF, the solution should satisfy an additional condition, $\gamma_{b2r,i} \geq \gamma^*$ for $i = 1, 2, \ldots, N$.

RS Selection

The RS selection has a strong effect on the network performance, since it determines the link conditions of both the transmissions between the MS and the RS and that between the RS and the BS. Each connection has a number of potential RSs, which can be near the source MS (for the uplink) or destination MS (for the downlink). For the uplink, the transmissions start from the source MSs, which assume that cooperative relaying does not exist. The objective of these transmissions is to achieve the target SINR γ^* at the BS receiver for each connection through power control, which works in the same way as in the conventional CDMA networks without relaying. During this period, the potential RSs for connection i listen to the transmissions from MS i

and measure the received SINRs. Those potential RSs with received SINRs satisfying a certain condition report to the BS, and the one with the best link gain to the BS is selected as the RS for the connection. When using DF, only the potential RSs that can correctly decode the signal from their source MS report to the BS. Although correct decoding is not a requirement for AF, we set an SINR threshold, $\gamma_{th} > 0$, so that only the potential RSs that receive the source MS's signal with an SINR above this threshold report to the BS. This is to avoid the case when the RS amplifies most interference and noise instead of the desired signal. The effect of this threshold on the system performance will be demonstrated later on. The RS selection process in the downlink works in a similar way. Initially, the BS transmits to the destination MSs, assuming there is no relaying. The transmission power to each destination MS is to achieve the target SINR for their connections. During this period, the potential RSs measure received SINRs and report to the BS. This process is the same as in the uplink. Upon receiving the reports from the potential RSs, the BS selects the one with the best link gain to the destination MS as the RS and notifies the RS through the downlink transmissions. In case there is no potential RS satisfying the required SINR condition for a connection, the source MS (for the uplink) or the BS (for the downlink) transmits in both the odd and even time slots.

4.2.3 Performance Results

We consider a single cell network, where the BS is located at the center of the cell. The initial locations of the MSs are uniformly distributed in the cell coverage area. MSs move randomly using a random way point model. We randomly select N MSs to carry N connections, and the remaining MSs can be potential RSs. Link gains include distance-based path loss and log-normal shadowing. The link gain for each channel is kept constant for a period of time, which follows the Gaussian distribution with a mean of 40 time slots and a standard deviation of 4 time slots, and then is regenerated. Default parameters are listed in Table 4.1.

When using either the optimum or heuristic power distributions, if there is no feasible solution, a simple removal scheme is used by removing the connection with the worst link gain from the source to the destination. The removed connection is in outage. This process is repeated until a feasible solution can be found for all the remaining connections. Below we examine both the outage probability and transmission power performance. Figs. 4.13-4.15 show the performance in the uplink, and Figs. 4.16-4.20 show the performance in the downlink.

Given a certain outage probability requirement, Fig. 4.13 shows the average transmission power, which is the total transmission power from all the source MSs and all the RSs divided by the number of the connections. For comparison, performance of the system without relaying is also shown. Since the transmission time of the source MSs is doubled in the system without relaying, the instantaneous transmission rate is $R/2$ in order to have the same average transmission rate as the system with relaying. It can be seen that using cooperative relaying can significantly reduce

Table 4.1 Default parameters settings

Parameter	Value
Cell size	2km
Spread spectrum bandwidth W	5MHz
MS velocity	0 to 20m/s
Path loss exponent	4
Standard deviation of log-normal fading	6dB
Background noise power η	10^{-8}W
Total number of MSs	140
Max. transmission power of MSs $P_{MS,max}$	0.5W
Max. transmission power of BSs $P_{BS,max}$	5W
SINR threshold γ^*	7dB
AF SINR threshold for RS selection, γ_{th}	4dB
Number of connections N	10
Data rate R	32kbps

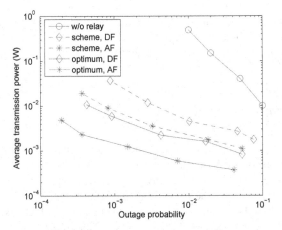

Fig. 4.13 Uplink: transmission power allocation vs. outage probability

the transmission power. Although the number of transmitters in the system with re-laying is increased, the total transmission power can be several magnitudes lower than that without relaying. As the outage probability becomes lower, the difference between required transmission power in the systems with and without cooperative relaying becomes larger. This can be a very attractive feature for encouraging mobile users to cooperatively relay traffic for each other. For a long term, every station may have its own traffic. Relaying traffic for each other should benefit all the users for reducing communication outage probability and saving battery power.

When the number of potential RSs that report to the BS is large, signaling trans-missions from the potential RSs to the BS can consume a large amount of the net-work resources. In order to limit the overhead for RS selection, the number of po-tential RSs for each connection should be limited. On the other hand, having a small number of potential RSs may not help improve the transmission performance. In

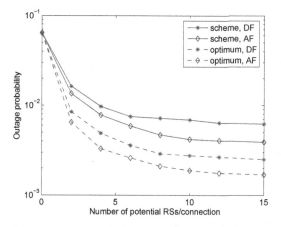

Fig. 4.14 Uplink: outage performance vs. number of potential RSs per connection

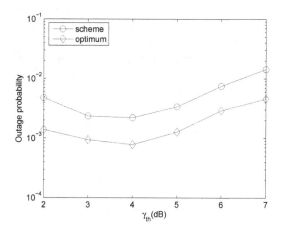

Fig. 4.15 Uplink: effect of γ_{th} on outage performance using AF

order to control the number of potential RSs, the source MS in the uplink or the destination MS in the downlink can send an RS request message, whose power depends on the number of potential RSs that the MS intends to have. Any MSs that do not carry their own traffic and hear the RS request message can be potential RSs of the requesting MS. The transmission power of the RS request message can be higher if the requesting MS intends to have more potential RSs, and the transmission power can be lower if a smaller number of potential RSs is sufficient. Fig. 4.14 shows the outage probability by varying the number of potential RSs per connection. For both the optimum and the heuristic power distributions, the outage performance is improved as there are more potential RSs. When the number of potential RSs per connection is relatively small, increasing the number can dramatically reduce the

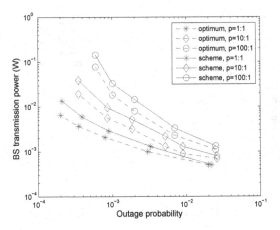

Fig. 4.16 Downlink: BS transmission power using AF

outage probability. However, when it exceeds a certain value, for example, 7 in the example shown, further increasing the number of potential RSs does not improve the outage probability, since the performance is then limited by other system parameters, such as the relative positions of the MSs and the BS.

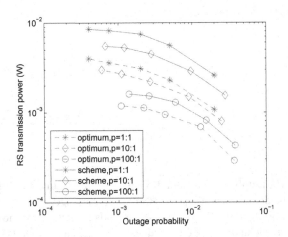

Fig. 4.17 Downlink: RS transmission power using AF

Fig. 4.15 shows the effect of γ_{th} on the outage performance in cooperative communications using AF. It is seen that for both the optimum and heuristic power distributions, there is an optimum value of γ_{th} that results in the lowest outage performance. This is due to the contradictory effects of γ_{th} in AF. When γ_{th} is very small, the link condition from the source MS to a selected RS may be poor. The poor receiving quality at the RS is then translated to poorer SINR after the RS forwards the amplified signal (together with interference and noise) to the BS. On the other hand, a large value of γ_{th} may prevent a "good" RS from being selected, and the connection cannot take advantage of the cooperative relaying. The optimum value of γ_{th} is between zero and γ^* in general and around 4dB in the example shown.

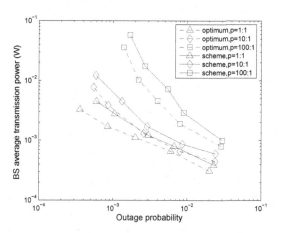

Fig. 4.18 Downlink: BS transmission power using DF

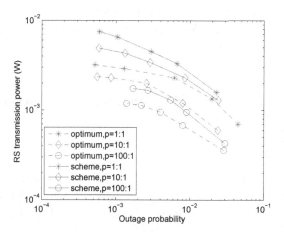

Fig. 4.19 Downlink: RS transmission power using DF

We then consider the performance in the downlink. Figs. 4.16 and 4.17, respectively, show the BS and RS transmission power when using AF; and Figs. 4.18 and 4.19, respectively, show the BS and RS transmission power when using DF. We

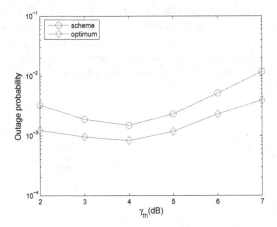

Fig. 4.20 Downlink: effect of γ_{th} on outage performance using AF

define $p = P_{BS,max}/P_{MS,max}$. When $P_{BS,max}$ is fixed, adjusting p in the simulation changes $P_{MS,max}$. For both DF and AF, the transmission power using the heuristic scheme is slightly higher than the optimum due to the fact that the heuristic scheme assigns the target SINRs to the BS-to-destination and RS-to-destination hops based on link gains without considering the respective interference conditions. It is seen that for both AF and DF, when the allowed RS transmission power increases, i.e., p is smaller, the required BS transmission power is reduced in order to achieve the same outage performance, or the outage probability can be reduced for the same maximum BS transmission power. Fig. 4.20 shows the effect of selecting different γ_{th} values in AF on the downlink outage probability, where we have similar observations as in the uplink.

4.3 Cognitive Radio Networks with Spectrum Underlay

4.3.1 System Description

Cognitive radio (CR) has been emerging as a result of two contradictory phenomena, the scarcity of available radio spectrum to satisfy increasing demands for wireless communications and the low utilization in the current licensed spectrum. The main reason for such dilemma is the fixed radio spectrum allocations. CR is a key technology in order to resolve such a problem by building secondary networks in the current licensed spectrum. The low utilization of the licensed spectrum provides opportunities to serve traffic in the secondary networks with a certain QoS requirement. Building CR networks (CRNs) in the licensed spectrum not only saves the high cost for accessing the licensed spectrum, but also has advantages over the networks working in the license-free spectrum, where effective QoS provisioning and resource management is very difficult due to the increasingly crowded spectrum usage. There are mainly two approaches for a CR (or secondary) device to accessing the licensed spectrum: spectrum overlay and spectrum underlay. In spectrum overlay, a CR device opportunistically uses those bands that are currently not used by licensed (or primary) devices for transmissions. In spectrum underlay, the CR devices can transmit simultaneously with the primary devices, as long as the interference level caused by the secondary transmissions in the primary network is below a certain level. A general definition for the interference temperature model is defined by FCC in [11].

For spectrum underlay, the interference at the primary receivers is measured and compared with an interference threshold. The comparison results are fed back to the CRN transmitters and help them make transmission decisions. Consider that the primary network is a cellular network and uses different frequency bands for the uplink and the downlink, and the CRN accesses only the uplink band of the primary network, then the interference experienced by the primary BSs determine whether the secondary nodes can transmit and how much power they can transmit. A controller for the CRN can be co-located with the primary BS for measuring the interference level at the BS. The controller can be a device independent of the primary BS. Alternatively, the primary and secondary networks can be tightly coupled, and the primary BS can work as the controller for the CRN and provide information to assist the secondary network transmissions.

A typical application for the tightly coupled case is that some MSs in a cellular network may decide to communicate directly with each other through their ad hoc air interface, instead of communicating with the BS. This can happen when the MSs are close to each other, for example, for a group activity in a small playground area. In this case, using direct communications requires lower transmission power than using the cellular mode, and the battery energy can be saved. In addition, the direct transmissions between the MSs without involving the BS can potentially save the communication cost of the users. On the other hand, such transmissions should not violate the QoS of other users in the cellular mode. Since the one-hop transmission range in an ad hoc network is usually much shorter than the cellular transmission range, the transmission power in the CRN on average is much less than in the cellular network, and the secondary-to-primary (s2p) interference is low in general, which makes it possible to support high transmission rates in the CRN. Due to its secondary nature, the CRN does not have control over the interference from the primary transmissions, which is referred to as primary-to-secondary (p2s) interference. The mutual interference between the two networks is illustrated in Fig. 4.21. The rest of this section is mainly based on the work in [9]. The primary network is a generic cell of a CDMA-based cellular network, which uses different frequency bands for the uplink and downlink transmissions, and the CRN accesses the uplink channel of the primary network. Power distributions in both the primary and secondary networks are jointly considered, the optimum transmission rates in the secondary network are solved based on different objectives, and the interference threshold is studied at the end.

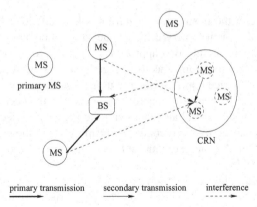

primary transmission   secondary transmission   interference

Fig. 4.21 Mutual interference between a primary cellular network and an ad hoc CRN

4.3.2 Joint Power Distributions

There can be two types of interference models for regulating the secondary transmissions. First, the s2p interference is measured at the BS receiver of a primary link and compared with an interference threshold. In this case, interference from the secondary transmitters should be separated from the interference from the primary transmitters to calculate the s2p interference level. This is difficult in the considered scenario, where both the primary and secondary links can transmit at the same time and may use the same waveform format. The second interference model is that for each received signal at the BS receiver, the sum of the noise, s2p interference, and the interference from the primary transmitters is measured. In this way, the s2p interference is not detected separately from the total interference. This model does not require a priori knowledge of the RF environment, consequently does not need to distinguish licensed signals from interference and noise, and this is the interference model considered below.

Denote the transmission power of a primary transmitter as $P_{p,i}$, for $i = 1, 2, \ldots, M_p$, and that of a secondary transmitter as $P_{s,i}$, for $i = 1, 2, \ldots, M_s$, where M_p and M_s, respectively, are the total number of primary and secondary links (or MSs). The transmission rate of the ith primary transmitter is $R_{p,i}$, and that of a secondary transmitter is $R_{s,i}$. Let $g_{p2b,i}$ denote the link gain from the transmitter of the ith primary link to the BS receiver, $g_{p2s,ij}$ the link gain from the transmitter of the ith primary link to the receiver of the jth secondary link, $g_{s2b,j}$ the link gain from the transmitter of the jth secondary link to the BS receiver, and $g_{s2s,ij}$ the link gain from the transmitter of the ith secondary link to the receiver of the jth secondary link. Each link has a strict SINR requirement, which should be above γ_p^* for a primary link and γ_s^* for a secondary link.

Consider the ith primary link, its required SINR can be satisfied if

$$\frac{G_{p,i}P_{p,i}g_{p2b,i}}{\sum_{j=1,j\neq i}^{M_p}P_{p,j}g_{p2b,j}+\sum_{j=1}^{M_s}P_{s,j}g_{s2b,j}+\eta} \geq \gamma_p^*, \tag{4.47}$$

where $G_{p,i} \doteq W/R_{p,i}$ is the processing gain of the ith primary link. In the denominator on the left-hand side of (4.47), the first term is the interference from all other primary links, the second term is the interference from all the secondary links, and η is the power of the background noise.

The ith secondary link experiences interference from all other secondary links and all primary links, and its SINR can be satisfied if

$$\frac{WP_{s,i}g_{s2s,ii}}{R_{s,i}\left(\sum_{j=1,j\neq i}^{M_s}P_{s,j}g_{s2s,ji}+\sum_{j=1}^{M_p}P_{p,j}g_{p2s,ji}+\eta\right)} \geq \gamma_s^*. \tag{4.48}$$

The transmission power of the secondary links is also limited by the interference threshold, I_{th}. That is,

$$\sum_{j=1,j\neq i}^{M_p}P_{p,j}g_{p2b,j}+\sum_{j=1}^{M_s}P_{s,j}g_{s2b,j}+\eta \leq I_{th}, \tag{4.49}$$

for $i = 1,2,\ldots,M_p$. That is, the total experienced noise and interference at the receiver of a primary link cannot exceed I_{th}. Note that condition (4.49) only limits the transmission of the secondary links. Occasionally, (4.49) may not be satisfied even if there is no secondary transmission. In this case, no secondary link is allowed to transmit, and the primary network will apply procedures, such as admission control or connection removal, to coordinate the transmissions of the primary links, but this is not related to the primary-secondary scenario.

4.3.3 Optimum Transmission Rate

Based on the above discussion we formulate the secondary link rate allocation problem as the following optimization problem,

$$\max_{\mathbf{R}_s,\mathbf{P}_s,\mathbf{P}_p} f(\mathbf{R}_s) \tag{4.50}$$

$$\text{s.t.} \quad \frac{WP_{s,i}g_{s2s,ii}}{R_{s,i}\left(\sum_{j=1,j\neq i}^{M_s}P_{s,j}g_{s2s,ji}+\sum_{j=1}^{M_p}P_{p,j}g_{p2s,ji}+\eta\right)} \geq \gamma_s^*, \; i = 1,2,\ldots,M_s \tag{4.51}$$

$$\frac{G_{p,i}P_{p,i}g_{p2b,i}}{\sum_{j=1,j\neq i}^{M_p}P_{p,j}g_{p2b,j}+\sum_{j=1}^{M_s}P_{s,j}g_{s2b,j}+\eta} \geq \gamma_p^*, \; i = 1,2,\ldots,M_p \tag{4.52}$$

$$\sum_{j=1,j\neq i}^{M_p}P_{p,j}g_{p2b,j}+\sum_{j=1}^{M_s}P_{s,j}g_{s2b,j}+\eta \leq I_{th}, \; i = 1,2,\ldots,M_p \tag{4.53}$$

$$0 \leq P_{p,i} \leq P_{p,\max}, \; i = 1,2,\ldots,M_p, \tag{4.54}$$

$$0 \leq P_{s,i} \leq P_{s,\max}, \; i = 1,2,\ldots,M_s, \tag{4.55}$$

which maximizes a certain function of the secondary link transmission rates, subject to the SINR requirements of the primary and secondary links and the given interference threshold, where $\mathbf{R}_s = (R_{s,1}, R_{s,2}, \ldots, R_{s,M_s})$, $\mathbf{P}_s = (P_{s,1}, P_{s,2}, \ldots, P_{s,M_s})$, and $\mathbf{P}_p = (P_{p,1}, P_{p,2}, \ldots, P_{p,M_p})$. We consider two objective functions based on different rate allocation criteria, a simple equal rate allocation (ERA) and a proportional fair rate allocation (PRA). With ERA, all secondary links transmit at the same rate. In this case, $R_{s,i} = R_s$ for all $i = 1, 2, \ldots, M_s$, and the objective of the optimization is to maximize R_s, i.e., $f(\mathbf{R}_s) = R_s$. The ERA formulation provides perfect fairness among all secondary links. However, links with poor SINR conditions transmit higher power in order to achieve the same rate as other links. Therefore, the link with the worst link condition limits the transmission rate of all links in the secondary network. In contrast, PRA is based on the proportional fair rate allocation [12], and the objective function is $f(\mathbf{R}_s) = \max \sum_{i=1}^{M_s} \log(R_{s,i})$. Because of the logarithm function, when $R_{s,i}$ is greater than some large value, further increasing it has little effect on the objective function. Therefore, when the service rates for the links with good channel conditions are sufficiently high, the remaining resources can be used by other links to further increase the objective function. In this sense, proportional fairness is a good tradeoff between resource utilization and users' satisfaction.

It can be found that the above optimization problem is a geometric programming (GP) problem, which is non-linear and non-convex due to constraints (4.51) and (4.52). By observing the two constraints we find that the left-hand side of each of the inequalities is the division of a monomial to a posynomial and is usually named as an inverted posynomial [10]. Lower bounding an inverted posynomial is allowed in GP since it is equivalent to upper bounding a posynomial, thus the problem can be easily transformed into a standard form GP problem. By applying the convex form transformation described in Appendix B (Case 1), this problem can be converted to a convex optimization problem and solved efficiently.

4.3.4 Interference Threshold

With a larger interference threshold, higher transmission power is allowed from the secondary links. This causes higher interference to the primary links. As a result, transmission power of the primary users will increase. This mutual interference effect eventually reaches a balance, provided there is a feasible solution to the problem formulated in the previous subsection. At this point, neither the primary nor the secondary links can increase their transmission power, and the interference level at the primary receiver is maximized, which can be equal to or less than I_{th}. Theoretically, the maximum interference level is limited by both the maximum transmission power of the primary transmitters, $P_{p,\max}$, and the maximum transmission power of the secondary transmitters, $P_{s,\max}$. The former is related to the amount of interference that the primary receivers can tolerate, and the latter determines the level of the interference that can be generated. When $P_{s,\max}$ is sufficiently large, the chance that a secondary transmitter can transmit at the maximum power is little, because the interference at the primary receiver would reach the interference threshold before any of the secondary transmitters reach the maximum transmission power. In such a scenario, the transmission power of the secondary transmitters is dependent on the co-channel interference within the secondary network and the mutual interference between the two networks, and is limited by the the interference threshold. When I_{th} is too large, the secondary nature of the CRN does not exist, which can negatively affect performance in the primary network. Intuitively, I_{th} should be set to be sufficiently large to allow more transmission opportunities in the CRN. Meanwhile, it should be below some value so that to protect the QoS in the primary network. Therefore, it is important to analyze and find an upper bound of I_{th}, given the QoS requirements in the primary network.

Consider that homogeneous traffic is carried out by the primary links. Then $G_{p,i} = G_p$ for all $i = 1, 2, \ldots, M_p$. Let I_p represent the aggregate noise and interference from all other primary links and all secondary transmitters that a particular primary link experiences at the BS receiver input. With perfect power control, the actual SINR for the primary link at the BS receiver input is equal to γ_p^*, and all the primary links have an equal received power, S_p, at the BS. In this case, we have $G_p S_p / I_p = \gamma_p^*$, or $I_p = G_p S_p / \gamma_p^*$. As the transmission power is limited by $P_{p,\max}$, we have

$$S_p = P_{p,i} g_{p2b,i} \leq P_{p,\max} g_{p2b,i}. \tag{4.56}$$

Then

$$I_p \leq \frac{G_p P_{p,\max} g_{p2b,i}}{\gamma_p^*}, \tag{4.57}$$

for all $i = 1, 2, \ldots, M_p$. Then the maximum interference level at the BS receiver should be

$$I_{p,\max} = \min_i \frac{G_p P_{p,\max} g_{p2b,i}}{\gamma_p^*}, \tag{4.58}$$

which is the upper bound of the interference threshold in the sense that the actual interference level at the primary link receiver can never be larger than $I_{p,\max}$, regardless how much the interference threshold is. Otherwise, transmissions from the secondary network can violate the QoS in the primary network.

Since $g_{p2b,i}$'s are random, below we find the distribution of $I_{p,\max}$ and then its mean value. Given any value $y > 0$,

$$\Pr\{I_{p,\max} \leq y\} = \Pr\left\{\min_i \frac{G_p P_{p,\max} g_{p2b,i}}{\gamma_p^*} \leq y\right\} \tag{4.59}$$

$$= \Pr\left\{\min_i g_{p2b,i} \leq \frac{y \gamma_p^*}{G_p P_{p,\max}}\right\} \tag{4.60}$$

$$= 1 - \prod_i \Pr\left\{g_{p2b,i} > \frac{y \gamma_p^*}{G_p P_{p,\max}}\right\} \tag{4.61}$$

$$= 1 - \left[1 - \Pr\left\{g_{p2b,i} \leq \frac{y \gamma_p^*}{G_p P_{p,\max}}\right\}\right]^{M_p}, \tag{4.62}$$

where $\Pr\left\{g_{p2b,i} \leq \frac{y\gamma_p^*}{G_pP_{p,\max}}\right\}$ depends on specific distribution of the link gain from the primary MS to the BS. In this derivation we assume that all $g_{p2b,i}$'s are independent and identically distributed. The mean of $I_{p,\max}$ can be found as

$$E[I_{p,\max}] = \int_0^\infty (1 - \Pr\{I_{p,\max} \leq y\})dy. \tag{4.63}$$

We find $\Pr\left\{g_{p2b,i} \leq \frac{y\gamma_p^*}{G_pP_{p,\max}}\right\}$ for a special case. Consider that the transmitted signals suffer from both path loss and log-normal fading. Then $g_{p2b,i} = Ad_{p2b,i}^{-\alpha}e^{-\beta X}$, where A is the link gain at a reference distance, $d_{p2b,i}$ is the distance (normalized to the reference distance) between the transmitter of the ith primary link to the BS, α is the path loss exponent, $\beta = \ln 10/10$, and X is a normally distributed random variable with zero mean and a standard deviation of σ. The distribution of $g_{p2b,i}$ can be found as

$$\Pr\left\{g_{p2b,i} \leq \frac{y\gamma_p^*}{G_pP_{p,\max}}\right\}$$

$$= \Pr\left\{Ad_{p2b,i}^{-\alpha}e^{-\beta X} \leq \frac{y\gamma_p^*}{G_pP_{p,\max}}\right\}$$

$$= \Pr\left\{X \geq \frac{-\alpha\ln d_{p2b,i} - \ln\left(\frac{y\gamma_p^*}{AG_pP_{p,\max}}\right)}{\beta}\right\}$$

$$= \int_0^D Q\left(-\frac{\alpha\ln z + \ln\left(\frac{y\gamma_p^*}{AG_pP_{p,\max}}\right)}{\beta\sigma}\right) f_d(z)dz, \tag{4.64}$$

where $Q(x) = \frac{1}{2\pi}\int_x^\infty e^{-\frac{u^2}{2}}du$, and $f_d(z)$ is the probability density function (pdf) of $d_{p2b,i}$. When all MSs are uniformly distributed in a circular area of radius of D, $f_d(z) = 2z/D^2$ for $0 \leq z \leq D$.

4.3.5 Performance Results

We consider a generic cell in a CDMA-based cellular network as the primary network. A number of MSs are uniformly distributed in the circular coverage area, where the BS is located at the center. Among all the MSs, M_p MSs are randomly selected as the primary transmitters for communicating directly with the BS in the uplink, and M_s MSs are then randomly selected from the remaining MSs as the secondary transmitters. For each secondary transmitter, a receiver is randomly selected from the remaining MSs within the ad hoc coverage of the transmitter. Default simulation parameters are listed in Table 4.2.

Table 4.2 Default simulation parameters

Parameter	Value
Transmission rate of each primary link R_p	64kbps
SINR threshold $\gamma_p^* = \gamma_s^*$	5dB
Path loss exponent α	4
Standard deviation of log-normal fading σ	4dB
Maximum MS transmission power $P_{p,\max}$ and $P_{s,\max}$	0.5W
Spread spectrum bandwidth W	5MHz
Background noise power η	10^{-10}W
Radius of BS coverage D	1000m
Maximum ad hoc transmission distance of MSs	300m

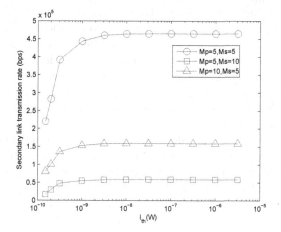

Fig. 4.22 Secondary link transmission rate: ERA

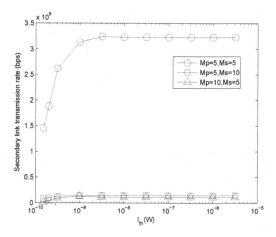

Fig. 4.23 Secondary link transmission rate: PRA

Figs. 4.22 and 4.23 show that when the interference threshold is below a certain value, the secondary link transmission rate increases with the interference threshold for both ERA and PRA. Beyond this range, further increasing the interference threshold does not affect the secondary transmission rate anymore. This is due to the mutual interference between the primary and secondary links. When the interference threshold is relatively low, little interference can be tolerated by the primary links, and the secondary links can only transmit very low power. Within this range, a higher interference threshold allows higher transmission power from the secondary links, but the secondary transmission power is still sufficiently low and does not cause significant interference to the primary links. Therefore, the interference threshold limits the capacity in the secondary network, and the secondary link transmission rate increases with the interference threshold. As the interference threshold further increases, the secondary links are allowed to transmit higher power, which causes higher interference to the primary links and increases their transmission power. Eventually, the transmission power is limited by the maximum transmission power of the transmitters, and the capacity in the secondary network is no longer affected by the interference threshold. Further increasing the interference threshold does not help the secondary links transmit higher power, and the secondary link rate is kept constant.

Figs. 4.22 and 4.23 also show that increasing the number of the primary or secondary links results in lower secondary link rate due to that more links are competing for the network resources. Comparing the two figures we can find that using PRA can achieve a lot higher transmission rate for the secondary links than using ERA, since the former can take better advantage of good channel conditions by allowing users with good link conditions to transmit at hither rates.

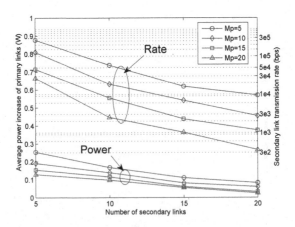

Fig. 4.24 Average power increase of primary links: ERA

With the presence of secondary transmissions, transmission power of the primary links is increased due to extra interference. Fig. 4.24 shows the average power

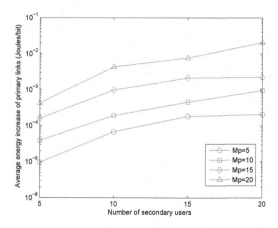

Fig. 4.25 Average energy increase of primary links: ERA

increase (API) of the primary links when ERA is used in the secondary network, where API is defined as the difference of the total amount of primary user transmission power with and without the secondary transmissions divided by the number of secondary links. The figure shows that API decreases with the number of secondary links. This is explained by the dramatic decrease of the secondary link transmission rate as the number of secondary links increases. Because of the equal rate allocation, the transmission rate of the secondary links is limited by the one with the worst link condition. As the number of links increases, the worst link gain becomes worse, and the service rate decreases. As a result, the transmission power of the secondary transmitters decreases, causing less interference to the primary links. Dividing the API by the total transmission rate of the secondary links, we have the average energy increase (AEI) per bit in Fig. 4.25, which shows that as the number of secondary links increases, it costs more energy from the primary transmitters on average for every transmitted bit in the secondary network.

For PRA, Figs. 4.26 and 4.27 show that both API and AEI of the primary links increase with the number of secondary links. Although having more secondary links increases the mutual interference, the total transmission rate of the secondary links using PRA is much higher than that using ERA. However, comparing Fig. 4.27 with Fig. 4.25 we find that the API in PRA is about 4 magnitudes lower than that in ERA. Equivalently, with the same amount of API, PRA can achieve much higher transmission rate than ERA. This shows that using PRA can much better utilize the resources available for the CRN.

Assume I_{th} is unlimited and QoS of the primary links is guaranteed, Figs. 4.28-4.29 show the average interference level at the BS receiver input versus different system parameters. The mean of the actual interference level at the BS is also shown for ERA and PRA, respectively. It is seen that a primary network with a smaller cell size or larger fading deviation can experience higher interference for the same number of secondary links. These relationships can be easily seen from (4.58). Most

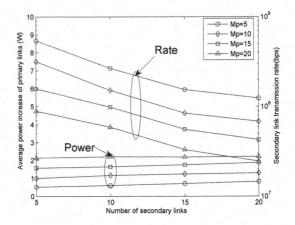

Fig. 4.26 Average power increase of primary links: PRA

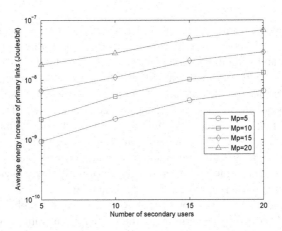

Fig. 4.27 Average energy increase of primary links: PRA

importantly, the figures indicate that using PRA results in significantly lower inter-ference to the primary links than using ERA. With ERA, the link with the worst SINR condition may transmit much higher power than other links. When this link is close to the primary link receiver (the BS), the primary transmitters should trans-mit very high power to combat the high interference. Therefore, in ERA, it is more likely that the primary link transmitters reach the maximum transmission power limit. This also explains that the interference level at the BS using ERA is almost the same as the calculated bound. In contrast, PRA does not encourage secondary links with poor SINR to transmit at high rate, and therefore, the interference level at the primary links is much lower.

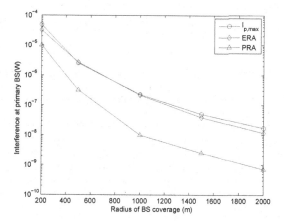

Fig. 4.28 Maximum s2p interference vs. cell size

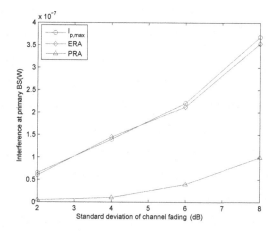

Fig. 4.29 Maximum interference vs. standard deviation of channel fading σ

References

1. Qiao C, Wu H (2000) ICAR: An integrated cellular and ad-hoc relay system. Proc. Ninth International Conference on Computer Communications and Networks: 154-161.
2. Lin YD, Hsu YC (2000) Multihop cellular: a new architecture for wireless communications. Proc. IEEE INFOCOM 3: 1273-1282.
3. Jiang H, Wang P, Zhuang W, Shen X (2007) An interference aware distributed resource management scheme for CDMA-based wireless mesh backbone. IEEE Transactions on Wireless Communications, 6(12): 4558-4567.
4. Zhao D, Todd TD (2006) Cellular CDMA capacity with out-of-band multihop relaying. IEEE Transactions on Mobile Computing 5(2): 170-178.
5. Rouse T, Band I, McLaughlin S (2002) Capacity and power investigation of opportunity driver multiple access (odma) networks in TDD-CDMA based systems. IEEE International Confer-

ence on Communications.
6. Harrold TJ, Nix AR (2002) Performance analysis of intelligent relaying in UTRA TDD. IEEE Vehicular Technology Conference: 13741378.
7. Wang B, Zhao D (2008) Optimum power distribution for uplink channel in a cooperative wireless CDMA network. IEEE International Conference on Communications (ICC).
8. Wang B, Zhao D (2009) Downlink power distribution of a wireless CDMA network with cooperative relaying. IEEE International Conference on Communications (ICC).
9. Wang B, Zhao D (2008) Performance analysis in CDMA-based cognitive wireless networks with spectrum underlay. Proc. IEEE Globecom.
10. Chiang M, Tan CW, Palomar DP, O'Neill D, Julian D (2007) Power control by geometric programming. IEEE Transactions on Wireless Communications 6(7): 2640-2651.
11. FCC (2003) Establishment of interference temperature metric to quantify and manage interference and to expand available unlicensed operation in certain fixed mobile and satellite frequency bands. ET Docket 03-289, Notice of Inquiry and Proposed Rulemaking.
12. Kelly FP, Maulloo AK, Tan DKH (1998) Rate control in communication networks: shadow prices, proportional fairness and stability. Journal of Operational Research Society 49: 237-252.

Appendix A
Irreducible Matrix and Dominant Eigenvalue

Definition: An $n \times n$ matrix $\mathbf{A}$ is irreducible if there is no permutation of coordinates such that

$$\mathbf{P}^T \mathbf{A} \mathbf{P} = \begin{bmatrix} \mathbf{A}_{11} & \mathbf{A}_{12} \\ \mathbf{0} & \mathbf{A}_{22} \end{bmatrix} \tag{A.1}$$

where $\mathbf{P}$ is an $n \times n$ permutation matrix with each row and each column having exactly one element of 1 and all other elements of 0, $\mathbf{A}_{11}$ is $r \times r$, $\mathbf{A}_{22}$ is $(n-r) \times (n-r)$, and $\mathbf{A}_{12}$ is $n \times (n-r)$. That is, an irreducible matrix cannot be placed into block upper-triangular form by simultaneous row/column permutations.

Theorem: A nonnegative $n \times n$ matrix $\mathbf{A}$ is irreducible if and only if $(\mathbf{I}+\mathbf{A})^{n-1} \succ \mathbf{0}$, where $\mathbf{I}$ is an $n \times n$ identity matrix, and $\succ$ is element-wise larger than.

Definition: Let λ_i, $i = 1, 2, \ldots, n$, be the eigenvalues of an $n \times n$ matrix $\mathbf{A}$. Then the spectral radius of the matrix is defined as $\rho(\mathbf{A}) \stackrel{\text{def}}{=} \max_i(|\lambda_i|)$.

Perron-Frobenius theorem for irreducible matrices: If $\mathbf{A} = (a_{ij})$ is an $n \times n$ nonnegative and irreducible matrix, then

- one of its eigenvalues is positive and greater than or equal to (in absolute value) all other eigenvalues. Such an eigenvalue is called the "dominant eigenvalue" or Perron-Frobenius eigenvalue of the matrix;
- there is a positive eigenvector corresponding to that eigenvalue; and
- $\rho(\mathbf{A})$ is equal to the dominant eigenvalue of the matrix and satisfies

$$\min_i \sum_j a_{ij} \leq \rho(\mathbf{A}) \leq \max_i \sum_j a_{ij}.$$

References

1. Varga RS (1962) Matrix iterative analysis, Chapter 2, Prentice-Hall, Inc., Englewood Cliffs, N.J.

Appendix B
Posynomial and Related Optimization Problems

Definition: A monomial is a function of the form

$$h(\mathbf{x}) = d x_1^{a^{(1)}} x_2^{a^{(2)}} \ldots x_n^{a^{(n)}}, \tag{B.1}$$

where d is nonnegative, x_i's are positive real numbers, and $a^{(i)}$'s are real numbers. Monomials are closed under multiplication and division.

Definition: A posynomial is a sum of monomials and of the form

$$f(\mathbf{x}) = \sum_{k=1}^{K} d_k x_1^{a_k^{(1)}} x_2^{a_k^{(2)}} \ldots x_n^{a_k^{(n)}}, \tag{B.2}$$

where d_k's are nonnegative, x_i's are positive real numbers, and $a_k^{(i)}$'s are real numbers. Posynomials are closed under addition, multiplication, and nonnegative scaling.

Definition: A standard geometric programming (GP) problem is as follows

$$\min f_0(\mathbf{x}) \tag{B.3}$$
$$\text{s.t. } f_i(\mathbf{x}) \leq 1, \ i = 1, 2, \ldots, m \tag{B.4}$$
$$h_l(\mathbf{x}) = 1, \ l = 1, 2, \ldots, n \tag{B.5}$$

where $f_i(\mathbf{x})$, $i = 0, 1, \ldots, m$, are posynomials,

$$f_i(\mathbf{x}) = d_{ik} x_1^{a_{ik}^{(1)}} x_2^{a_{ik}^{(2)}} \ldots x_n^{a_{ik}^{(n)}}, \tag{B.6}$$

and $h_l(\mathbf{x})$, $l = 1, 2, \ldots, n$, are monomials.

Consider the following optimization problem

$$\min f_0(\mathbf{x}) \tag{B.7}$$
$$\text{s.t. } f_i(\mathbf{x}) \leq 1, \ i = 1, 2, \ldots, m \tag{B.8}$$

Case 1: When $f_i(\mathbf{x})$ for $i = 0, 1, \ldots, m$ are all posynomials of the form (B.6), the problem is a standard GP problem, which in general is not convex, but can be transformed into a convex problem. With a change of variables: $y_i = \log x_i$ and $b_{ik} = \log d_{ik}$, (B.7) can be converted into convex form [2]:

$$\min \tilde{f}_0(\mathbf{y}) = \log \left(\sum_k e^{\mathbf{a}_{0k}^T \mathbf{y} + b_{0k}} \right) \tag{B.9}$$

$$\text{s.t. } \tilde{f}_i(\mathbf{y}) = \log \left(\sum_k e^{\mathbf{a}_{ik}^T \mathbf{y} + b_{ik}} \right) \leq 0, \ i = 1, 2, \ldots, m, \tag{B.10}$$

where $\mathbf{a}_{ik} = (a_{ik}^{(1)}, a_{ik}^{(2)}, \ldots, a_{ik}^{(n)})^T$. Since the functions $\tilde{f}_i(\mathbf{x})$ are convex, this problem is a convex optimization problem, which can be solved globally and efficiently through the interior point primal dual method [2] with polynomial running time.

Case 2: When $f_0(\mathbf{x})$ is convex, and $f_i(\mathbf{x})$, $1 \leq i \leq m$, is in the format of a ratio of posynomials, i.e., $f_i(\mathbf{x}) = s(\mathbf{x})/g(\mathbf{x})$, the optimization problem is not convex and difficult to solve directly. A successive approximation method is designed in [1], where the basic idea is to solve such a problem by a series of approximations, each of which can be optimally solved in an easy way.

The problem can be turned into a geometrical programming (GP) problem by approximating the denominator of the ratio of posynomials, $g(\mathbf{x})$, with a monomial $\tilde{g}(\mathbf{x})$, but leaving the numerator $s(\mathbf{x})$ unchanged. It is proved in [1] that if $g(\mathbf{x}) = \sum_i u_i(\mathbf{x})$ is a posynomial, then

$$g(\mathbf{x}) \geq \tilde{g}(\mathbf{x}) = \prod_i \left[\frac{u_i(\mathbf{x})}{\alpha_i} \right]^{\alpha_i}. \tag{B.11}$$

If, in addition, $\alpha_i = \frac{u_i(\mathbf{x}_0)}{g(\mathbf{x}_0)}$, $\forall i$, for any fixed positive $\mathbf{x}_0$, then $\tilde{g}(\mathbf{x}_0) = g(\mathbf{x}_0)$, and $\tilde{g}(\mathbf{x}_0)$ is the best local monomial approximation to $g(\mathbf{x}_0)$ near $\mathbf{x}_0$ in the sense of the first order Taylor approximation. It is further proved in [1] that the approximation of a ratio of posynomials $f_i(\mathbf{x}) = s(\mathbf{x})/g(\mathbf{x})$ with $\tilde{f}_i(\mathbf{x}) = s(\mathbf{x})/\tilde{g}(\mathbf{x})$ satisfies the Karush-Kuhn-Tucker (KKT) conditions:

(1) $f_i(\mathbf{x}) \leq \tilde{f}_i(\mathbf{x})$ for all $\mathbf{x}$,
(2) $f_i(\mathbf{x}_0) = \tilde{f}_i(\mathbf{x}_0)$ where $\mathbf{x}_0$ is the optimal solution of the approximated problem in the previous iteration, and
(3) $\nabla f_i(\mathbf{x}_0) = \nabla \tilde{f}_i(\mathbf{x}_0)$.

With the above process, the denominator of $f_i(\mathbf{x})$ is approximated as a monomial, and $f_i(\mathbf{x})$ is then approximated as a posynomial. An iterative method as follows is then proposed in [1] to solve the original optimization problem:

Step 0: Choose an initial feasible point $\mathbf{x}^{(0)}$ and set $j = 1$.
Step 1: Approximate $g_i(x)$ with $\tilde{g}_i(\mathbf{x})$ around the previous point $\mathbf{x}^{(j-1)}$.
Step 2: Solve the approximated problem and obtain solution $\mathbf{x}^{(j)}$.
Step 3: Increase j by 1 and go back to Step 2 until the solution converges.

The convergency of this method is guaranteed by the KKT conditions in the approximation.

References

1. Chiang M (2005) Geometric programming for communication systems. Foundations and Trends in Communications and Information Theory 2(1-2): 1-154.
2. Boyd S, Vandenberghe L (2004) Convex Optimization. Cambridge University Press.